7 771424 14

D1549699

Solving Dynamics Problems in MATLAB

Brian D. Harper
Mechanical Engineering
The Ohio State University

A supplement to accompany
Engineering Mechanics: Dynamics, 5th Edition
by J.L. Meriam and L.G. Kraige

North Lanarkshire Council Motherwell Library Hamilton Road, Motherwell CUM	
7 771424 14	
Askews	18-Dec-2002
620.104	£18.95
	1195946

JOHN WILEY & SONS, INC.

Cover Photo: Courtesy NASA

To order books or for customer service call 1-800-CALL-WILEY (225-5945).

Copyright © 2002 John Wiley & Sons, Inc. All rights reserved.

No part of this publication may be reproduced, stored in a retrieval system or transmitted in any form or by any means, electronic, mechanical, photocopying, recording, scanning or otherwise, except as permitted under Sections 107 or 108 of the 1976 United States Copyright Act, without either the prior written permission of the Publisher, or authorization through payment of the appropriate per-copy fee to the Copyright Clearance Center, 222 Rosewood Drive, Danvers, MA 01923, (978) 750-8400, fax (978) 750-4470. Requests to the Publisher for permission should be addressed to the Permissions Department, John Wiley & Sons, Inc., 605 Third Avenue, New York, NY 10158-0012, (212) 850-6011, fax (212) 850-6008, E-Mail: PERMREQ@WILEY.COM.

ISBN 0-471-20311-4

Printed in the United States of America

10 9 8 7 6 5 4 3 2 1

Printed and bound by Hamilton Printing Company

CONTENTS

Introduction **5**

Chapter 1 An Introduction to MATLAB **7**

Numerical Calculations 7
Writing Scripts (m-files) 10
Defining Functions 12
Graphics 13
Symbolic Calculations 21
Differentiation and Integration 24
Solving Equations 26

Chapter 2 Kinematics of Particles **37**

2.1 Sample Problem 2/4 (Rectilinear Motion) 38
2.2 Problem 2/94 (Rectangular Coordinates) 41
2.3 Problem 2/120 (n-t Coordinates) 44
2.4 Sample Problem 2/10 (Polar Coordinates) 45
2.5 Problem 2/180 (Space Curvilinear Motion) 49
2.6 Sample Problem 2/15 (Constrained Motion of Connected Particles) 51

Chapter 3 Kinetics of Particles **55**

3.1 Sample Problem 3/3 (Rectilinear Motion) 56
3.2 Problem 3/92 (Curvilinear Motion) 59
3.3 Sample Problem 3/16 (Potential Energy) 61
3.4 Problem 3/208 (Linear Impulse/Momentum) 64
3.5 Problem 3/243 (Angular Impulse/Momentum) 66
3.6 Problem 3/358 (Curvilinear Motion) 67

Chapter 4 Kinetics of Systems of Particles **71**

4.1 Problem 4/25 (Conservation of Momentum) 72
4.2 Problem 4/58 (Steady Mass Flow) 74
4.3 Problem 4/82 (Variable Mass) 77

Chapter 5 Plane Kinematics of Rigid Bodies 81

5.1 Problem 5/12 (Rotation) 82
5.2 Problem 5/39 (Absolute Motion) 85
5.3 Sample Problem 5/9 (Relative Velocity) 87
5.4 Problem 5/110 (Instantaneous Center) 91
5.5 Sample Problem 5/15 (Absolute Motion) 93

Chapter 6 Plane Kinetics of Rigid Bodies 99

6.1 Problem 6/29 (Translation) 100
6.2 Sample Problem 6/4 (Fixed-Axis Rotation) 103
6.3 Problem 6/94 (General Plane Motion) 105
6.4 Problem 6/99 (General Plane Motion) 108
6.5 Sample Problem 6/10 (Work and Energy) 110
6.6 Problem 6/204 (Impulse/Momentum) 115

Chapter 7 Introduction to Three-Dimensional Dynamics of Rigid Bodies 119

7.1 Problem 7/27 (Rotation about a Fixed Point) 120
7.2 Sample Problem 7/3 (General Motion) 122
7.3 Sample Problem 7/6 (Kinetic Energy) 124

Chapter 8 Vibration and Time Response 129

8.1 Sample Problem 8/2 (Free Vibration of Particles) 130
8.2 Sample Problem 8/6 (Forced Vibration of Particles) 132
8.3 Problem 8/137 (Forced Vibration of Particles) 135
8.4 Problem 8/91 (Rigid Bodies) 137

INTRODUCTION

Computers and software have had a tremendous impact upon engineering education over the past several years and most engineering schools now incorporate computational software such as MATLAB in their curriculum. Since you have this supplement the chances are pretty good that you are already aware of this and will have to learn to use MATLAB as part of a Dynamics course. The purpose of this supplement is to help you do just that.

There seems to be some disagreement among engineering educators regarding how computers should be used in an engineering course such as Dynamics. I will use this as an opportunity to give my own philosophy along with a little advice. In trying to master the fundamentals of Dynamics there is no substitute for hard work. The old fashioned taking of pencil to paper, drawing free body and mass acceleration diagrams, struggling with equations of motion and kinematic relations, etc. is still essential to grasping the fundamentals of Dynamics. A sophisticated computational program is not going to help you to understand the fundamentals. For this reason, my advice is to use the computer only when required to do so. Most of your homework can and should be done without a computer.

The problems in this booklet are based upon problems taken from your text. The problems are slightly modified since most of the problems in your book do not require a computer for the reasons discussed in the last paragraph. One of the most important uses of the computer in studying Mechanics is the convenience and relative simplicity of conducting parametric studies. A parametric study seeks to understand the effect of one or more variables (parameters) upon a general solution. This is in contrast to a typical homework problem where you generally want to find one solution to a problem under some specified conditions. For example, in a typical homework problem you might be asked something about the trajectory of a particle launched at an angle of 30 degrees from the horizontal with an initial speed of 30 ft/sec. In a parametric study of the same problem you might typically find the trajectory as a function of two parameters, the launch angle θ and initial speed v. You might then be asked to plot the trajectory for different launch angles and speeds. A plot of this type is very beneficial in visualizing the general solution to a problem over a broad range of variables as opposed to a single case.

As you will see, it is not uncommon to find Mechanics problems that yield equations that cannot be solved exactly. These problems require a numerical approach that is greatly simplified by computational software such as MATLAB. Although numerical solutions are extremely easy to obtain in MATLAB this is still the method of last resort. Chapter 1 will illustrate several methods for obtaining symbolic (exact) solutions to problems. These methods should always be tried first. Only when these fail should you generate a numerical approximation.

Many students encounter some difficulties the first time they try to use a computer as an aid to solving a problem. In many cases they are expecting that they have to do something fundamentally different. It is very important to understand that there is no fundamental difference in the way that you would formulate computer problems as opposed to a regular homework problem. Each problem in this booklet has a problem formulation section prior to the solution. As you work through the problems be sure to note that there is nothing peculiar about the way the problems are formulated. You will see free-body and mass acceleration diagrams, kinematic equations etc. just like you would normally write. The main difference is that most of the problems will be parametric studies as discussed above. In a parametric study you will have at least one and possibly more parameters or variables that are left undefined during the formulation. For example, you might have a general angle θ as opposed to a specific angle of 20°. If it helps, you can "pretend" that the variable is some specific number while you are formulating a problem.

This supplement has eight chapters. The first chapter contains a brief introduction to MATLAB. If you already have some familiarity with MATLAB you can skip this chapter. Although the first chapter is relatively brief it does introduce all the methods that will be used later in the book and assumes no prior knowledge of MATLAB. Chapters 2 through 8 contain computer problems taken from chapters 2 through 8 of your textbook. Thus, if you would like to see some computer problems involving the kinetics of particles you can look at the problems in chapter 3 of this supplement. Each chapter will have a short introduction that summarizes the types of problems and computational methods used. This would be the ideal place to look if you are interested in finding examples of how to use specific functions, operations etc.

This supplement uses the student edition of MATLAB version 5.3. MATLAB is a registered trademark of The Mathworks, Inc., 24 Prime Park Way, Natick, Massachusetts, 01760.

AN INTRODUCTION TO MATLAB

1

This chapter provides an introduction to the MATLAB programming language. Although the chapter is introductory in nature it will cover everything needed to solve the computer problems in this booklet.

1.1 Numerical Calculations

When you open MATLAB you should see a prompt something like the following.

```
EDU»
```

To the right of this prompt you will write some sort of command. The following example assigns a value of 200 to the variable a. Simply type "a = 200" at the prompt and then press enter.

```
EDU» a=200
a =
  200
```

There are many situations where you may not want MATLAB to print the result of a command. To suppress the output simply type a semi-colon ";" after the command. For example, to suppress the output in the above case type "a = 200;" and the press enter. The basic mathematical operations of addition, subtraction, multiplication, division and raising to a power are accomplished exactly as you would expect if you are familiar with other programming languages, i.e. with the keys +, -, *, \, and ^. Here are a couple of examples.

```
EDU» c=2*5-120/4
c =
  -20
```

```
EDU» d=2^4
d =
   16
```

MATLAB has many built in functions that you can use in calculations. If you already know the name of the function you can simply type it in. If not, you can find a list of functions in the "Help Desk (HTML)" file available under Help in the main menu bar. Here are a few examples.

```
EDU» e=sin(.5)+tan(.2)
e =
   0.6821
```

As with most mathematical software packages, the default unit for angles will be radians.

```
EDU» f=sqrt(16)
f =
    4

EDU» g=log(5)
g =
   1.6094

EDU» h=log10(5)
h =
   0.6990
```

In the last two examples be careful to note that *log* is the natural logarithm. For a base 10 logarithm (commonly used in engineering) you need to use *log10*.

Range Variables

In the above examples we have seen many cases where a variable (or name) has been assigned a numerical value. There are many instances where we would like a single variable to take on a range of values rather than just a single value. Variables of this type are often referred to as range variables. For example, to have the variable x take on values between 0 and 3 with an increment of 0.25 we would type "x=0:0.25:3".

```
EDU» x=0:0.25:3
x =
  Columns 1 through 7
       0   0.2500   0.5000   0.7500   1.0000   1.2500   1.5000
  Columns 8 through 13
```

```
 1.7500  2.0000  2.2500  2.5000  2.7500  3.0000
```

Range variables are very convenient in that they allow us to evaluate an expression over a range with a single command. This is essentially equivalent to performing a loop without actually having to set up a loop structure. Here's an example.

```
EDU» x=0:0.25:3;
EDU» f=3*sin(x)-2*cos(x)
f =
  Columns 1 through 7
  -2.0000  -1.1956  -0.3169   0.5815   1.4438   2.2163   2.8510
  Columns 8 through 13
   3.3084   3.5602   3.5906   3.3977   2.9936   2.4033
```

Although range variables are very convenient and almost indispensable when making plots, they can also lead to considerable confusion for those new to MATLAB. To illustrate, suppose we wanted to compute a value y = 2*x-3*x^2 over a range of values of x.

```
EDU» x=0:0.5:3;
EDU» y = 2*x-3*x^2
??? Error using ==> ^
Matrix must be square.
```

What went wrong? We have referred to x as a range variable because this term is most descriptive of how we will actually use variables of this type in this manual. As the name implies, MATLAB is a program for doing matrix calculations. Thus, internally, x is really a row matrix and operations such as multiplication or division are taken, by default, to be matrix operations. Thus, MATLAB takes the operation x^2 to be matrix multiplication x*x. Since matrix multiplication of two row matrices doesn't make sense, we get an error message. In a certain sense, getting this error message is very fortunate since we are not actually interested in a matrix calculation here. What we want is to calculate a certain value y for each x in the range specified. This type of operation is referred to as term-by-term. For term-by-term operations we need to place a period "." in front of the operator. Thus, for term by term multiplication, division, and raising to a power we would write ".*", "./" and ".^" respectively. Since matrix and term-by-term addition or subtraction are equivalent, you do not need a period before + or -. The reason that our earlier calculations did not produce an error is that they all involved scalar quantities. To correct the error above we write.

```
EDU» x=0:0.5:3;
EDU» y = 2*x-3*x.^2
y =
```

0 0.2500 -1.0000 -3.7500 -8.0000 -13.7500 -21.0000

Note that we wrote "2*x" instead of "2.*x" since 2 is a scalar. If in doubt, it never hurts to include the period.

1.2 Writing Scripts (m-files)

If you have been working through the examples above you have probably noticed by now that you cannot modify a line once it has been entered into the worksheet. This makes it rather difficult to debug a program or run it again with a different set of parameters. For this reason, most of your work in MATLAB will involve writing short scripts or m-files. The term m-file derives from the fact that the programs are saved with a ".m" extension. An m-file is essentially just a list of commands that you would normally enter at a prompt in the worksheet. After saving the file you just enter the name of the file at the prompt after which MATLAB will execute all the lines in the file as if they had been entered. The chief advantage is that you can alter the script and run it as many times as you like.

To get started, select "*File...New...M-File*" from the main menu. A screen like that below will then pop up.

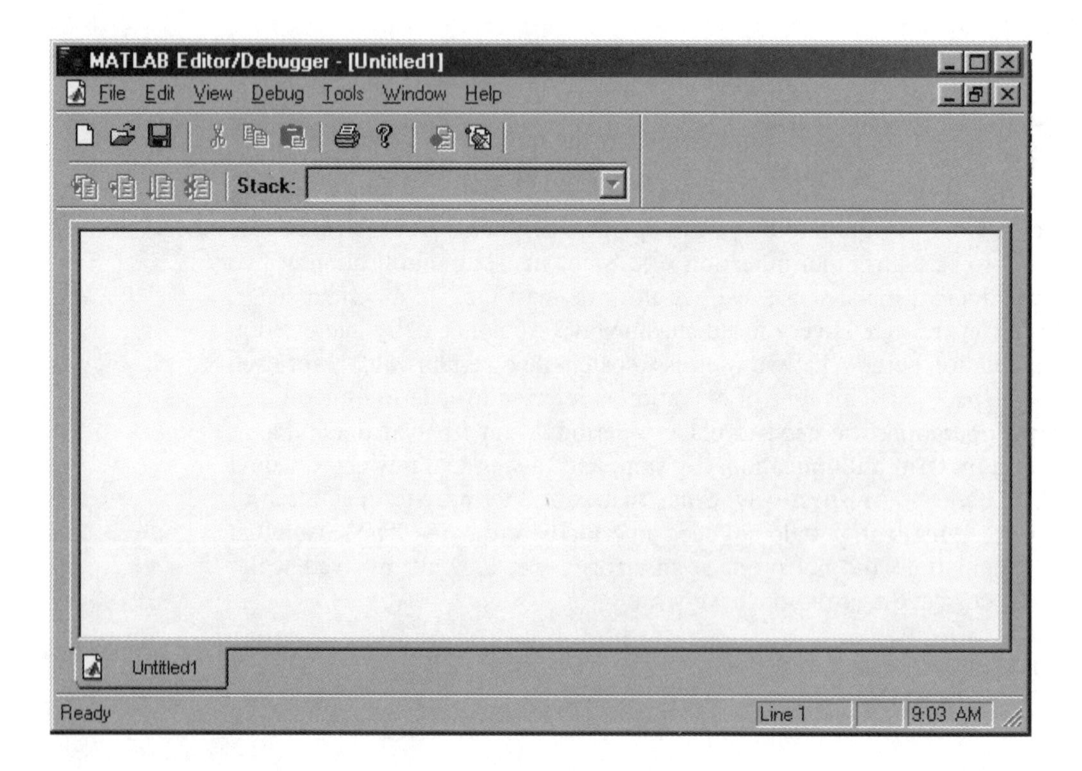

Now enter the program line by line into the editor. Basically, you just enter the commands you would normally enter at a prompt (do not, however, enter >> at the beginning of a line). If you want to enter comments, begin the line with "%". These lines will not be executed. Once you have finished entering the script into the editor you can save the file (be sure to remember where you saved it). The file name can contain numbers but cannot begin with a number. It also cannot contain any mathematical operators such as +, -, *, / etc. Also, the name of the file should not be the same as a variable it computes.

To run the script all you have to do is type the name of the program at a prompt (>>), but without the ".m" extension. A common error at this point is that MATLAB will give some response indicating that it hasn't a clue what you are doing. The reason this happens is that the file you saved is not in the current path. You can check (or change) the path by selecting *File...Set Path...* from the main menu.

Following is a simple example of a script file that sets up a range variable x and then calculates two functions over the range.

```
%%%%%%%%%%%%%%%%%%%%%%%%%%%%%%%%%%%%%%%%%%%
% This script calculates two
% functions over a specified range
x=0:0.2:1;
f=sin(x)./(1+x)
g=sin(x).*exp(x)
%%%%%%%%% end of script %%%%%%%%%%%%%%%%%%%%
```

The program was saved as example1, so here is the output you should see when the program is entered at a prompt.

```
EDU» example1
f=
     0   0.1656   0.2782   0.3529   0.3985   0.4207
g =
     0   0.2427   0.5809   1.0288   1.5965   2.2874
```

To save space we will, in the future, show the output of a script immediately after the script is given without actually showing the file name being entered at a prompt.

1.3 Defining Functions

Although MATLAB has many built in functions (see the "Help Desk (HTML)" file available under Help in the main menu bar), it is sometimes advantageous to define your own functions. While user defined functions are actually special types of m-files (they will be saved with the .m extension) their use is really quite different from the m-files discussed above. For example, suppose you were to create a function that you call f1. You would then use this function as you would any built in function like sin, cos, log etc.

Since functions are really special types of m-files you start by opening an editor window with "*File...New...M-File*". The structure of the file is, however, quite different from a script file. Following is an m-file for a function called load1. Note that you should save the file under the same name as the function. The file would thus show up as load1.m in whatever directory you save it. As with script files, you need to be sure the current path contains the directory in which the file was stored.

```
%%%%%% a function m-file %%%%%%%%%%%%%%%%
function y = load1(x)
% this file defines the function load1(x)
y = 2*(x-1)+3*(x-2).^2;
%%%%%% end of function m-file %%%%%%%%%%%
```

It is important to understand the structure of a function definition before you try to create some on your own. The name of this function is "load1" and will be used as you would any other function. y and x are local variables which MATLAB uses to calculate the function. There is nothing special about y and x, other variable names will work as well. Now, suppose you were to type load1(3) at a prompt or within a script file. MATLAB will calculate the expression for y substituting x = 3. This value of y will then be returned as the value for load1(3).

Here are a few examples using our function load1.

```
EDU» load1(3)
ans =
   7

EDU» f=load1(5)-load1(3)
f =
   28
```

```
EDU» load1(load1(3))
ans =
   87

EDU» x=1:0.2:2;
EDU» g=load1(x)
g =
   3.0000   2.3200   1.8800   1.6800   1.7200   2.0000
```

1.4 Graphics

One of the most useful things about a computational software package such as MATLAB is the ability to easily create graphs of functions. As we will see, these graphs allow one to gain a lot of insight into a problem by observing how a solution changes as some parameter (the magnitude of a load, an angle, a dimension etc.) is varied. This is so important that practically every problem in this supplement will contain at least one plot. By the time you have finished reading this supplement you should be very proficient at plotting in MATLAB. This section will introduce you to the basics of plotting in MATLAB.

MATLAB has the capability of creating a number of different types of graphs. Here we will consider only the X-Y plot. The most common and easiest way to generate a plot of a function is to use range variables. The following example will guide you through the basic procedure.

```
EDU» x=-3:0.1:3;
EDU» f=x.*exp(-x.^2);
EDU» plot(x,f)
```

The procedure is very simple. First define a range variable covering the range of the plot, then define the function to be plotted and issue the plot command. After typing in the above you should see a graph window pop up which looks something like the following figure.

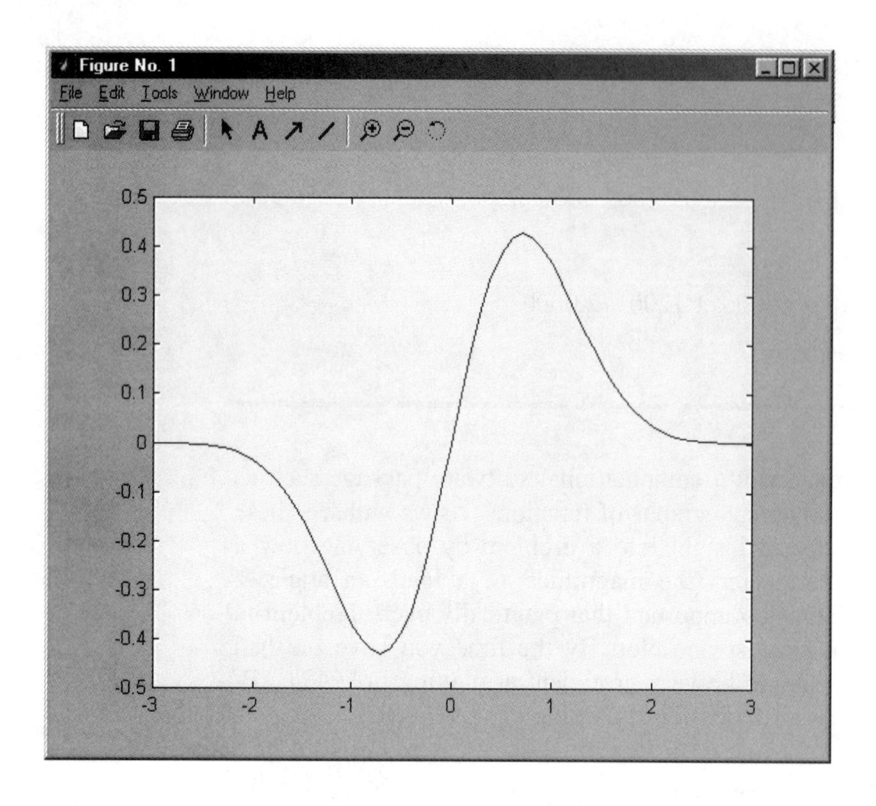

Things like titles and axis labels can be added in one of two ways. The first is by line commands such as those shown below.

```
EDU» xlabel('x')
EDU» ylabel('x*exp(-x^2)')
EDU» title('A Simple Plot')
```

Take a look at the graph window after you type each of the lines above. At this point, the graph window should look something like that shown below.

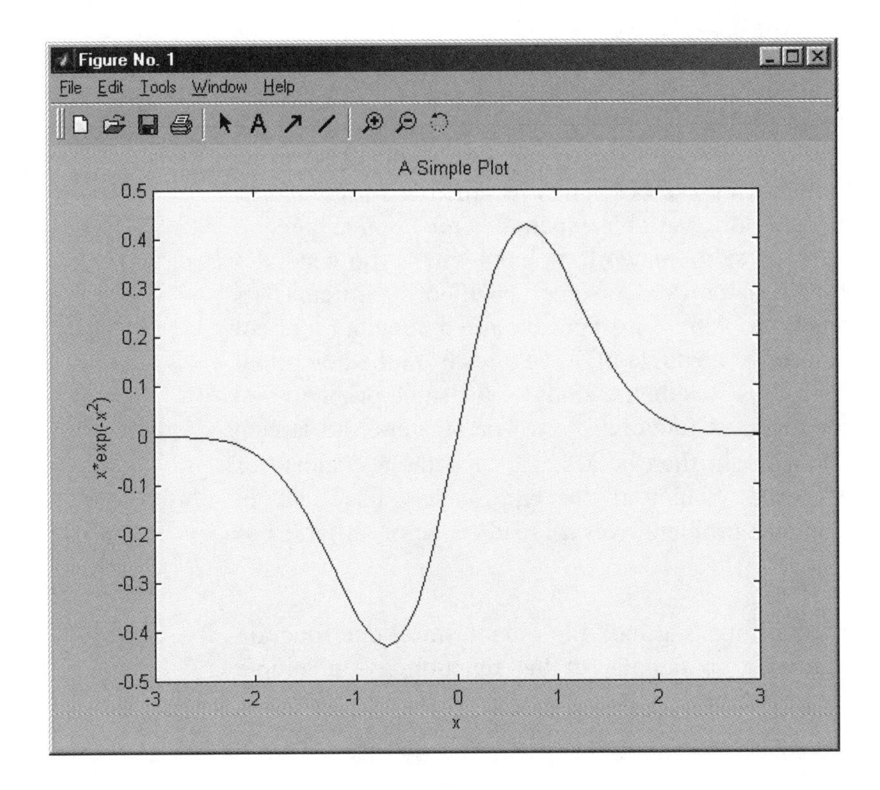

There are many other things that you can add or change on the plot. The easiest way to do this is with the second of the two approaches mentioned above. In the graph menu select "*Tools...Enable Plot Editing*". Now you can edit the plot in a number of different ways. To get a pop up menu allowing you to place a title and label the axes (such as was done with line commands above) simply double click on the graph. Alternately, you can right click on the graph and choose "*properties*" from the list. You can also add legends, text, lines etc. You should spend some time experimenting with a graph window to see the range of possibilities.

Once you have things like you want them you can save the graph to a file. You can also export the graph to some format suitable for inserting in, say, a word processor. Select "*File...Export*" in the graph window to see the range of possible formats.

Parametric Studies

One of the most important uses of the computer in studying Mechanics is the convenience and relative simplicity of conducting parametric studies (not to be confused with parametric plotting discussed below). A parametric study seeks to understand the effect of one or more variables (parameters) upon a general solution. This is in contrast to a typical homework problem where you generally want to find one solution to a problem under some specified conditions. For example, in a typical homework problem you might be asked something about the trajectory of a particle launched at an angle of 30 degrees from the horizontal with an initial speed of 30 ft/sec. In a parametric study of the same problem you might typically find the trajectory as a function of two parameters, the launch angle θ and initial speed v. You might then be asked to plot the trajectory for different launch angles and speeds. A plot of this type is very beneficial in visualizing the general solution to a problem over a broad range of variables as opposed to a single case.

Parametric studies generally require making multiple plots of the same function with different values of a particular parameter in the function. As a simple example, consider the following function.

$$f = 5 + x - 5x^2 + ax^3$$

What we would like to do is gain some understanding of how f varies with both x and a. It might be tempting to make a three dimensional plot in a case like this. Such a plot can, in some cases, be very useful. Usually, however, it is too difficult to interpret. This is illustrated by the following three dimensional plot of f versus x and a.

```
% This script produces a 3-d plot of
% f versus x and a.
x=linspace(-10,10,30);
a=linspace(-3,3,30);
[X,A]=meshgrid(x,a);
F=5+X-5*X.^2+A.*X.^3;
B=0.*A+2;
mesh(X,A,F)
colormap gray
%%%%%% end of script %%%%%%%%%%%%%%%%%%%%
```

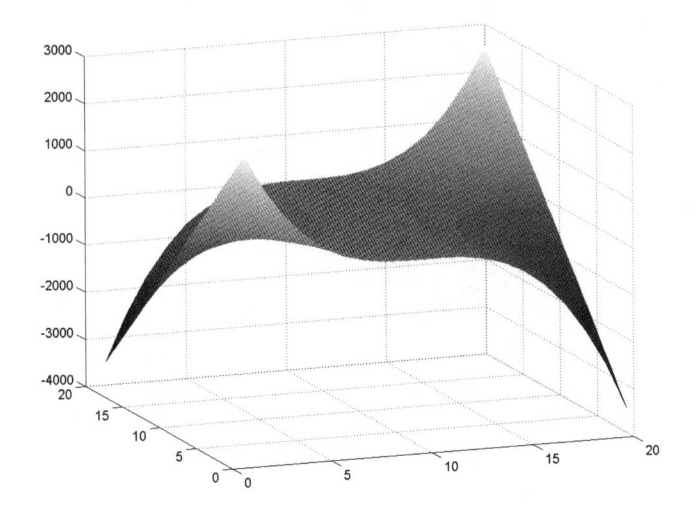

The plot above certainly is interesting but, as mentioned above, not very easy to interpret. In most cases it is much better to plot the function several times (with different values of the parameter of interest) on a single two-dimensional graph. We will illustrate this by plotting f as a function of x for a = -1, 0, and 1. This is accomplished in the following script.

```
% script for plotting f versus x for several
% values of a.
x=-5:0.1:5;
g=5+x-5*x.^2;
f1=g-x.^3; % a = -1
f2=g;      % a = 0
f3=g+x.^3; % a = 1
plot(x,f1,x,f2,x,f3)
xlabel('x')
ylabel('f')
%%%%%%%% end of script %%%%%%%%%%%%%%%%%%%%%%
```

Running this script results in the following plot.

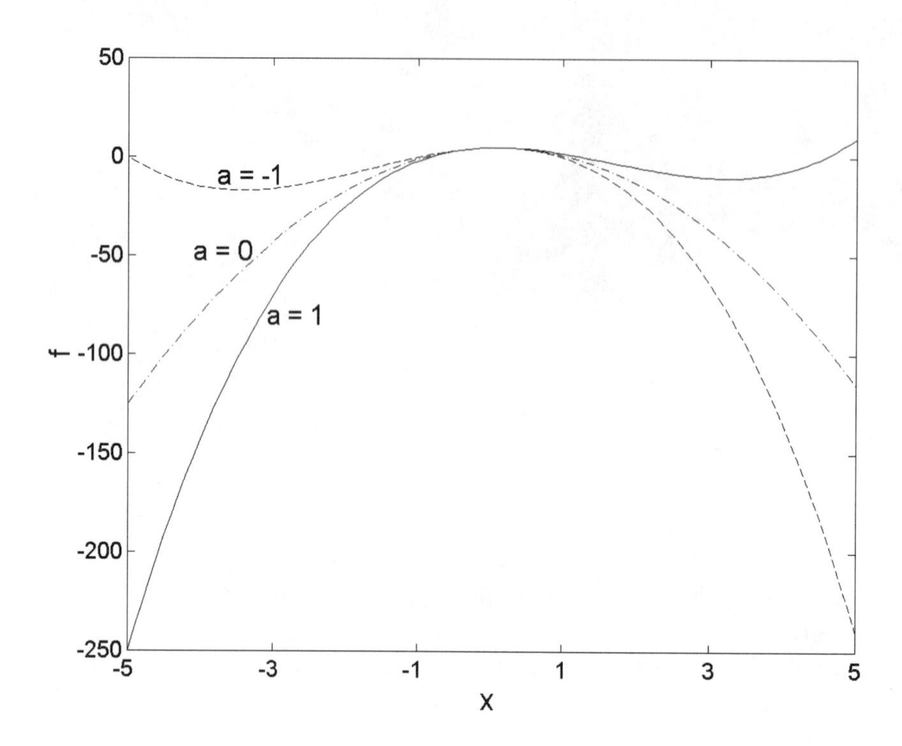

Parametric Plots

It often happens that one needs to plot some function y versus x but y is not known explicitly as a function of x. For example, suppose you know the x and y coordinates of a particle as a function of time but want to plot the trajectory of the particle, i.e. you want to plot the y coordinate of the particle versus the x coordinate. A plot of this type is generally called a parametric plot. Parametric plots are easy to obtain in MATLAB. You start by defining the two functions in terms of the common parameter and then define the common parameter as a range variable. Then issue a plot command using the two functions as arguments. The following script illustrates this procedure.

```
% This script file illustrates parametric plotting.
% A function f is plotted versus another function g.
% The two functions are related by the
% common parameter a.
a=-1:0.05:3.5;
f=10*a.*(2-a);
g=sin(3*a);
plot(g,f)
xlabel('g')
ylabel('f')
%%%%%%%% end of script %%%%%%%%%%%%%%%%%%%%%%%%%%
```

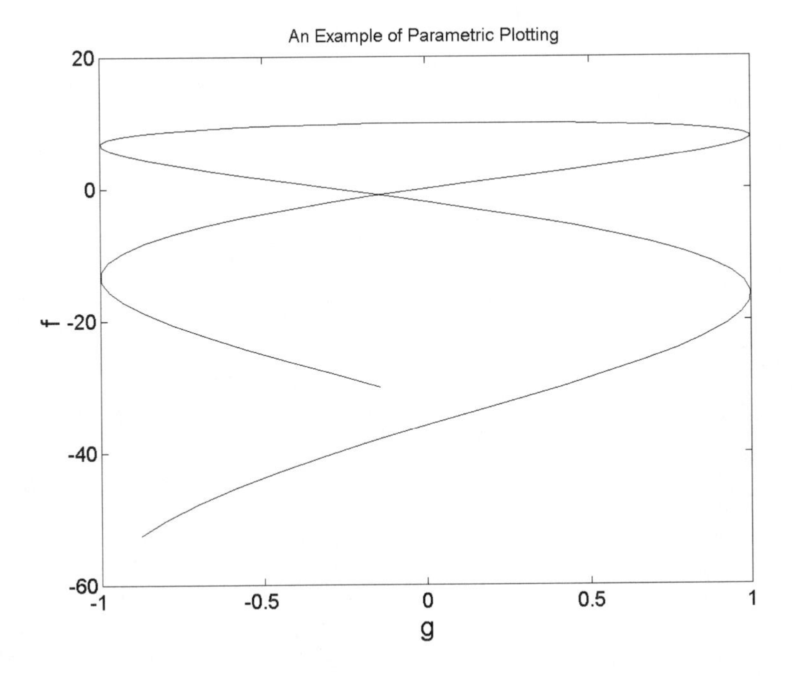

The range selected for the parameter can have a big, and sometimes surprising effect on the resulting graph. To illustrate, try increasing the upper limit on the range on a a few times and see how the graph changes.

You can, of course, also plot g as a function of f.

```
% Another example of parametric plotting
a=-1:0.05:6;
f=10*a.*(2-a);
g=sin(3*a);
plot(f,g)
xlabel('f')
ylabel('g')
%%%%%%%% end of script %%%%%%%%%%%%%%%%
```

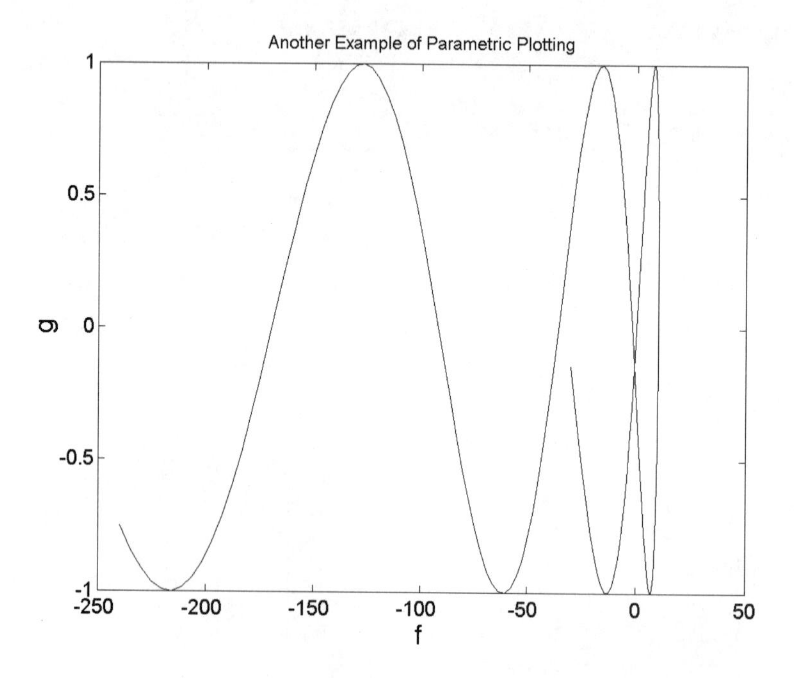

1.5 Symbolic Calculations

One of the most useful features of many modern mathematical software packages is the capability of doing symbolic math. By symbolic math we mean mathematical operations on expressions containing undefined variables rather than numbers. Maple is, perhaps, the best-known computer algebra program. Many other programs (including MATLAB) have a Maple engine that is called whenever symbolic operations are performed. The Maple engine is part of the Symbolic Math Toolbox so, if you do not have this toolbox, you will not be able to do symbolic math. If you have a student edition of MATLAB 5 you should have the Symbolic Math Toolbox. If you do not have the Symbolic Math Toolbox, you will need to perform whatever symbolic calculations are done in this manual by hand.

How does MATLAB decide whether to call Maple and attempt a symbolic calculation? Perhaps the best way to learn this is by studying the examples below and by practicing. Maple will always be used to manipulate any expression represented as a character string (i.e. an expression enclosed in 'quotation' marks). In the remainder of this manual we will refer to character strings of this type as Maple expressions. Here are a few examples of Maple expressions.

```
EDU» 'x+a*x^2-b*x^3'
ans =
x+a*x^2-b*x^3
EDU» f='A*sin(x)*exp(-x^2)'
f=
A*sin(x)*exp(-x^2)
```

Instead of placing the entire expression in quotations, you can also first declare certain variables to be symbols and then write the expressions as usual. This is more convenient in some situations but can get confusing. Here are some examples of this type.

```
EDU» syms x a b A           % Sets x, a, b and A as symbols
EDU» g=x+a*x^2-b*x^3
g=
x+a*x^2-b*x^3
EDU» f=A*sin(x)*exp(-x^2)
f=
A*sin(x)*exp(-x^2)
```

The above are just examples of Maple expressions and do not involve any symbolic math. Two of the most important applications of symbolic math will be discussed in the next two sections, namely symbolic calculus (integration and differentiation) and symbolic solution of one or more equations. The purpose of the present section is to introduce you to the basic procedures of symbolic math as well as to give a few other useful applications. One particularly useful symbolic operation is substitution and is illustrated by the following examples.

```
EDU» g='x+a*x^2-b*x^3'
g =
x+a*x^2-b*x^3

EDU» subs(g,'a',3)  % substitutes 3 for a in g. Note the quotation marks about a.
ans =
x+3*x^2-b*x^3

EDU» g1=subs(g,'x','(y-c)')
g1 =
(y-c)+a*(y-c)^2-b*(y-c)^3

EDU» syms x y A B
EDU» f=A*sin(x)*exp(-B*x^2)
f =
A*sin(x)*exp(-B*x^2)

EDU» subs(f,B,4)  % Note that B doesn't have to be in quotes since it
                    was declared to be a symbol.
ans =
A*sin(x)*exp(-4*x^2)
```

Another very convenient feature available in the Symbolic Toolbox provides the capability of defining "inline" functions thus avoiding the necessity for creating special m-files as described above. Here is an example.

```
EDU» f=inline('x^2*exp(-x)')
f =
   Inline function:
   f(x) = x^2*exp(-x)

EDU» f(3)
ans =
  0.4481
```

In the previous section we plotted the function $f = 5 + x - 5x^2 + ax^3$ for several values of a. It turns out to be somewhat more convenient to do this with an inline function. Let's see how it works.

```
EDU» f=inline('5+x-5*x^2+a*x^3')
f=
    Inline function:
    f(a,x) = 5+x-5*x^2+a*x^3
```

At this point we might anticipate a minor problem. When we plot this as a function of x we will have a range variable as one of the arguments for f. This creates a problem as MATLAB will be performing the numerical calculations and it will take the operations in f as matrix operations rather than term by term. Fortunately we can use the "*vectorize*" operator to insert periods in the appropriate places.

```
EDU» f=vectorize(f)
f=
    Inline function:
    f(a,x) = 5+x-5.*x.^2+a.*x.^3

EDU» x=-5:0.1:5;
EDU» plot(x,f1(-1,x),x,f1(0,x),x,f1(1,x))
```

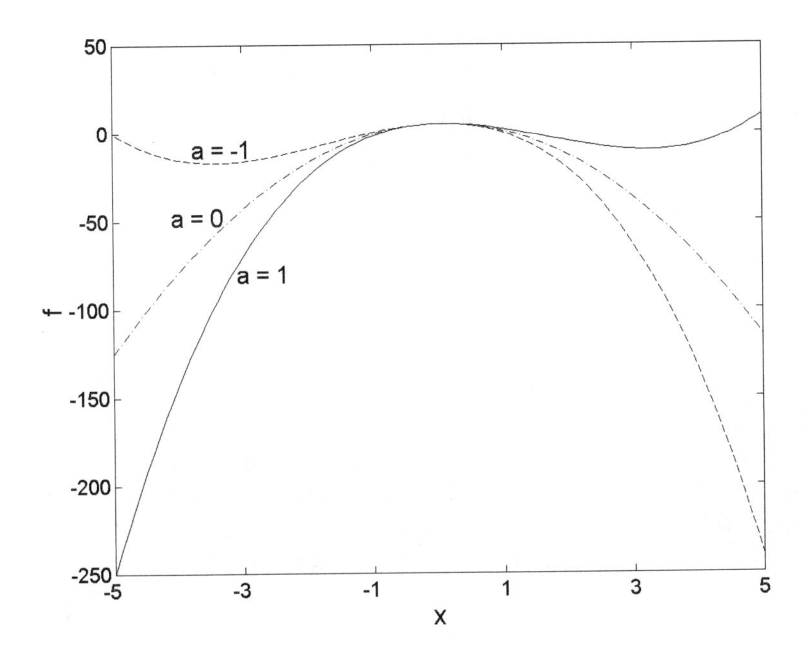

1.6 Differentiation and Integration

Here we will consider only symbolic differentiation and integration since these are most useful in the mechanics applications in this manual. For differentiation we will use the Maple *diff* command. Care should be exercised here since *diff* is also a MATLAB command. To guarantee that you are using the Maple command, be sure that you use *diff* only on Maple expressions.

```
EDU» f='a*sec(b*t)';
EDU» diff(f, 't')     % Note that t must be placed in quotation marks since
                        it is a symbol.
ans =
a*sec(b*t)*tan(b*t)*b
```

We can avoid using the quotation marks by making a, b and t symbols.

```
EDU» syms a b t
EDU» f=a*sec(b*t);
EDU» diff(f, t)
ans =
a*sec(b*t)*tan(b*t)*b
```

To find higher order derivatives use the command diff(f, x, n) where n is the order of the derivative. For example, to find the third derivative of a*log*(b+x) we would write

```
EDU» diff('a*log(b+x)', 'x', 3)
ans =
2*a/(b+x)^3
```

Symbolic integration will be performed with the *int* command. The general format of this command is int(f, x, a, b) where f is the integrand (a Maple expression), x is the integration variable and a and b are the limits of integration. If x, a and b have not been declared symbols with the *syms* command, they must be placed in quotation marks. If the integration limits are omitted, the indefinite integral will be evaluated. Here are several examples of definite and indefinite integrals.

```
EDU» int('sin(b*x)','x')
ans =
-1/b*cos(b*x)
```

```
EDU» int('sin(b*x)','x','c','d')
ans =
 -(cos(b*d)-cos(b*c))/b

EDU» syms x b c d
EDU» f=b*log(x);
EDU» int(f,x)
ans =
b*x*log(x)-b*x

EDU» int(f,x,c,d)
ans =
b*d*log(d)-b*d-b*c*log(c)+b*c
```

If a definite integral contains no unknown parameters either in the integrand or the integration limits, the *int* command will provide numerical answers. Here are a few examples.

```
EDU»  int('x+3*x^3','x',0,3)
ans =
261/4

EDU» int('log(x)','x',2,5)   % Don't forget that log is the natural logarithm.
ans =
5*log(5)-3-2*log(2)
```

Note that Maple will always try to return an exact answer. This usually results in answers containing fractions or functions as in the above examples. This is very useful in some situations; however, one often wants to know the numerical answer without having to evaluate a result such as the above with a calculator. To obtain numerical answers use Maple's *eval* function as in the following examples.

```
EDU» eval(int('x+3*x^3','x',0,3))
ans =
  65.2500
EDU» eval(int('log(x)','x',2,5))
ans =
   3.6609
```

1.7 Solving Equations

Our preferred approach to solving equations is to use Maple's *solve* command available in the *Symbolic Toolbox.* We will also discuss briefly MATLAB's built in function *fzero* which can be used to solve single equations numerically.

Solving single equations

Here are a couple of examples illustrating how to use Maple's *solve* command to solve a single equation symbolically. The first example is the (hopefully) familiar quadratic equation.

```
EDU» f='a*x^2+b*x+c=0';
EDU» solve(f, 'x')
ans =
[ 1/2/a*(-b+(b^2-4*a*c)^(1/2))]
[ 1/2/a*(-b-(b^2-4*a*c)^(1/2))]
```

```
EDU» g='a*sin(x)-cos(x)=1';
EDU» solve(g, 'x')
ans =
[                          pi]
[ atan2(2*a/(1+a^2),(a^2-1)/(1+a^2))]
```

Note that both examples produced two solutions with each solution being placed in brackets. One of the solutions to the second example is the number π. The second solution illustrates something you should always watch out for. In many cases you may get a solution in terms of a function that you are not familiar with, i.e. atan2 that appears in the second solution to the second example above. Whenever this happens you should type "help atan2" at a prompt in order to learn more about the function.

If the variable being solved for is the only unknown in the equation, *solve* will return a number as the result. Here are a couple of examples.

```
EDU» f='2*x^2+4*x-12';
EDU» solve(f, 'x')
ans =
[ -1+7^(1/2)]
[ -1-7^(1/2)]
```

```
EDU» g='5*sin(x)-cos(x)=1';
EDU» solve(g, 'x')
ans =
```

```
[        pi]
[ atan(5/12)]
```

If you want the answer as a floating point number instead of an exact result, use the Maple command *eval*. For example, with g defined as above we would write

```
EDU» eval(solve(g, 'x'))
ans =
   3.1416
   0.3948
```

Although Maple will always attempt to return an exact (symbolic) result, you may occasionally have a case where a floating point number is returned without using *eval*. Here is an example.

```
EDU» f='10*sin(z)=2*z^2+1';
EDU» solve(f, 'z')
ans =
.10227001429283204007554677150838
```

Whenever this occurs, Maple was unable to solve the equation exactly and switched automatically to a numerical solution. In a certain sense this is very convenient as the same command is being used for both symbolic and numerical calculations. The problem is that nonlinear equations quite often have more than one solution, as illustrated by the examples above. If you have the full version of Maple then you will have various options available for finding multiple numerical solutions. But since we are dealing now with numerical rather than symbolic solutions it makes more sense just to use MATLAB.

The appropriate MATLAB command for the numerical solution of an equation is *fzero*. This command numerically determines the point at which some defined function has a value of zero. This means that we need to rewrite our equation in terms of an expression whose zeros (roots) are solutions to the original expression. This is easily accomplished by rearranging the original equation into the form *expression* = 0. We will call this expression g in this example. Rewriting the above equation f we have $g = 2z^2 + 1 - 10\sin(z)$. The general format for fzero is fzero('function_name', x_0) where 'function_name' is the name of the m-file function that you have created and x_0 is an initial guess at the root (zero) of the function. The initial guess is very important since, if there is more than one solution, MATLAB will find the solution closest to the initial guess. Well, you may be wondering how we can determine an approximate solution if we do not yet know the solution. Actually, this is very easy to do. First we define a function g(z) whose roots will be the solution to the equation of interest. Next, we plot this function in order to estimate the location of points where g(z) = 0. Here's how it works for the present example.

```
%%%%% function m-file %%%%%%
function y = gz(z)
y = 2*z.^2 + 1 - 10*sin(z);
%%%%% end of m-file %%%%%%%%
```

Above is our user defined function (m-file) which we have named gz. Now we can plot this function over some range as illustrated below.

```
EDU» z=-1:0.01:3;
EDU» plot(z,gz(z),z,0) % we also plot a line at g=0 to help find the roots.
EDU» xlabel('z')
EDU» ylabel('g(z)')
```

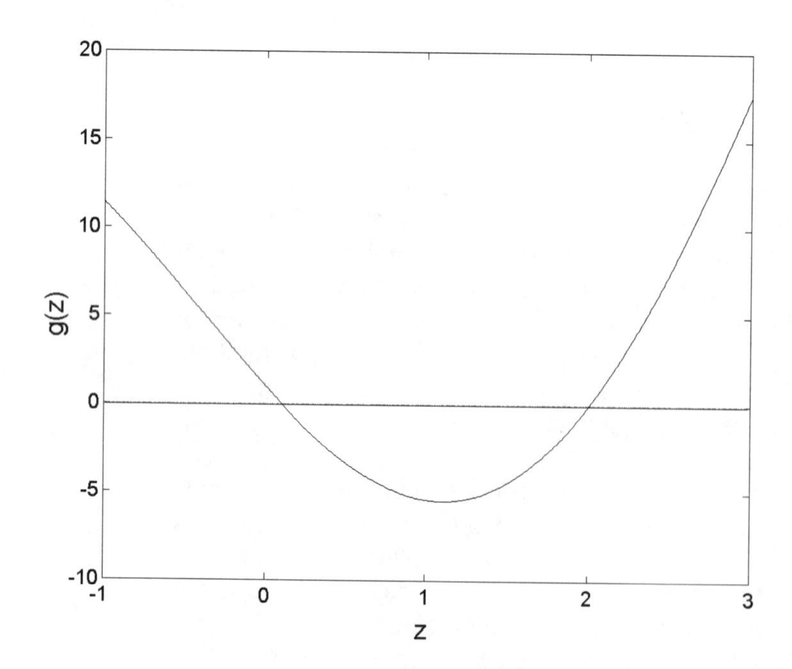

The plot shows that the function is zero at about z = 0 and 2. The first root is clearly that found by Maple's *solve* command above. Now we can easily find both roots with the *fzero* command.

```
EDU» fzero('gz', 0)
Zero found in the interval: [-0.11314, 0.11314].
ans =
   0.1023
```

EDU» fzero('gz', 2)
Zero found in the interval: [1.9434, 2.0566].
ans =
   2.0076

Thus, the two solutions are z = 0.1023 and 2.0076.

Finding Maxima and Minima of Functions

Here we will illustrate two methods for finding maxima and minima of a function.

Method 1. Using the symbolic diff and solve functions.

The usual method for finding maxima or minima of a function f(x) is to first determine the location(s) x at which maxima or minima occur by solving the equation $\frac{df}{dx}=0$ for x. One then substitutes the value(s) of x thus determined into f(x) to find the maximum or minimum. Consider finding the maximum and minimum values of the following function:

$$f = 1 + 2x - x^3$$

Before proceeding, it is a good idea to make a plot of the function f. This will give us a rough idea where the minima and maxima are as a check on the results we obtain below. The following script will plot f as a function of x.

```
% This script plots f as a function of x
x = -2:0.05:2;
f = 1 + 2*x - x.^3;
plot(x, f)
xlabel('x')
ylabel('f')
%%%%%%%%%% end of script %%%%%%%%%%%%
```

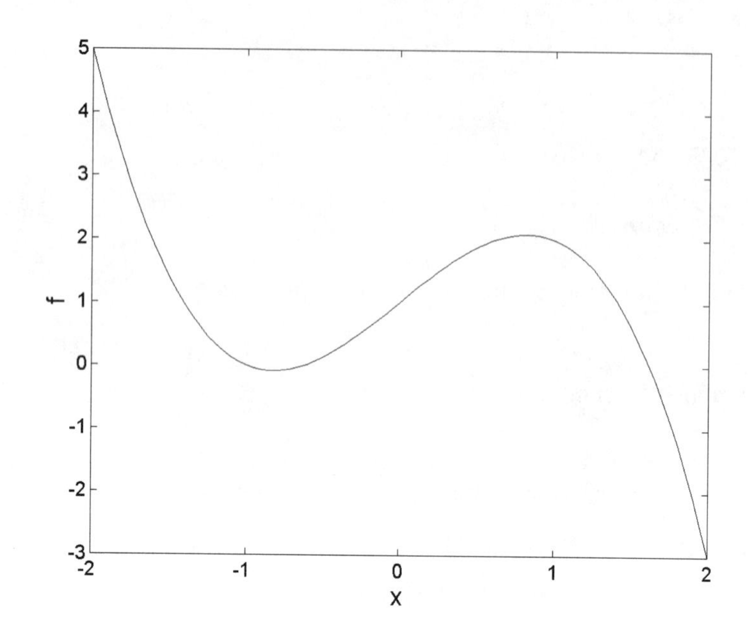

This plot shows we have a minimum of about 0 at x of about –0.9 and a maximum of about 2 at x of about 0.9. Now we can proceed to find the results more precisely.

```
EDU» f = '1+2*x-x^3'

f =
1+2*x-x^3

EDU» dfdx = diff(f,'x')
dfdx =
2-3*x^2

EDU» solve(dfdx,'x')
ans =
[ 1/3*6^(1/2)]
[ -1/3*6^(1/2)]
```

If an equals sign is omitted (as above) *solve* will find the root of the supplied Maple expression. In other words, it will solve the equation dfdx = 0. It is more convenient in the present case to have the solutions as floating point numbers. Thus, we should have used *eval* as in the following.

```
EDU» eval(solve(dfdx,'x'))
ans =
 0.8165
 -0.8165
```

The above results are the locations (x) where the minima and maxima occur. To find the maximum and minimum values of the function f we use subs to substitute these results back into f.

```
EDU» subs(f,'x',0.8165)
ans =
2.0887
```

```
EDU» subs(f,'x',-0.8165)
ans =
 -0.0887
```

Thus, $f_{max} = 2.0887$ at x = 0.8165 and $f_{min} = -0.0887$ at x = – 0.8165.

Method 2. Using MATLAB's min and max functions.

```
EDU» x = -2:0.05:2;
EDU» f = 1 + 2*x - x.^3;
```

These two lines are identical to those in the script used to plot f above. If you have just run that script you do not need to execute these lines. Now we can find the minimum and maximum values of f by typing *min(f)* and *max(f)*.

```
EDU» min(f)
ans = -3
```

```
EDU» max(f)
ans = 5
```

These are clearly not the answers we obtained above. What went wrong? It is important to understand that *min* and *max* do not find true minima and maxima in the mathematical sense, i.e. they do not find locations where df/dx = 0. Instead, they find the minimum or maximum value of all those calculated in the vector f. To see this, go back and look at the plot of f above and you will see that the minimum and maximum values of f are indeed – 3 and 5. This is another good reason to plot the function first. To avoid the above difficulty we can change the range of x to insure we pick up the true minima and maxima.

```
EDU» x = -1:0.05:1;
```

```
EDU» f = 1 + 2*x - x.^3;
```

In most cases, we want not only a minimum or maximum but also the location where they occur. This being the case we should use the following command.

```
EDU» [f_min,i] = min(f)
f_min = -0.0880
i = 5
```

The command [f_min, i] = min(f) finds the minimum value of f and assigns it to the variable f_min. The index of this value is assigned to the variable i. Now we can find the location of the minimum by printing the 5th value of the vector x.

```
EDU» x_min=x(i)
x_min = -0.8000
```

Now we can repeat the above for the maximum.

```
EDU» [f_max,i] = max(f)
f_max =  2.0880
i = 37

EDU» x_max=x(i)
x_max = 0.8000
```

You may have noticed at this point that the answers are somewhat different from those found by the first method above. The reason is, once again, that *min* and *max* do not find true minima and maxima. To obtain more accurate results you can use a closer spacing for the range variable (x). To see this, repeat the above for a spacing of 0.01 instead of 0.05.

Solving several equations simultaneously

```
EDU» eqn1='x^2+y^2=12'
eqn1 =
x^2+y^2=12
EDU» eqn2='x*y=4'
eqn2 =
x*y=4
```

In the above we have defined two equations which we will now solve for the two unknowns, x and y.

```
EDU» [x,y] = solve(eqn1,eqn2)
x =

[ 5^(1/2)-1]
[ -1-5^(1/2)]
[ 5^(1/2)+1]
[ 1-5^(1/2)]

y =

[ 5^(1/2)+1]
[ 1-5^(1/2)]
[ 5^(1/2)-1]
[ -1-5^(1/2)]
```

Note that four solutions have been found, the first being $x=\sqrt{5}-1$, $y=\sqrt{5}+1$. The following example illustrates a case where there are more unknowns than there are equations. In this case, the specified variables will be solved for in terms of the others.

```
EDU» f='x^2 + a*y^2 = 0'
f=
x^2 + a*y^2 = 0

EDU» g='x-y = b'
g =
x-y = b

EDU» [x,y] = solve(f,g)
x =

[ 1/2/(a+1)*(-2+2*(-a)^(1/2))*b+b]
[ 1/2/(a+1)*(-2-2*(-a)^(1/2))*b+b]

y =

[ 1/2/(a+1)*(-2+2*(-a)^(1/2))*b]
[ 1/2/(a+1)*(-2-2*(-a)^(1/2))*b]
```

In this case, there are two solutions. Now let's consider an example with three equations.

```
EDU» eqn1='x^2+y^2=12'
eqn1 =
x^2+y^2=12

EDU» eqn2='x*y=4'
eqn2 =
x*y=4

EDU» eqn3='x-y=z'
eqn3 =
x-y=z

EDU» [x,y,z] = solve(eqn1,eqn2,eqn3)
 x =

[  5^(1/2)+1]
[  1-5^(1/2)]
[  5^(1/2)-1]
[ -1-5^(1/2)]

 y =

[  5^(1/2)-1]
[ -1-5^(1/2)]
[  5^(1/2)+1]
[  1-5^(1/2)]

z =

[  2]
[  2]
[ -2]
[ -2]
```

There may be many cases where an exact (symbolic) solution cannot be found. In these cases, Maple will automatically make an attempt at a numerical solution. This is illustrated by the following example.

```
EDU» eqn1='sin(y)-exp(y)*x = 0'
eqn1 =
        sin(y)-exp(y)*x = 0

EDU» eqn2='x^2-y = 2'

eqn2 =
        x^2-y = 2

EDU» [x,y] = solve(eqn1,eqn2)

x =
         -1.1615309277258776360546221064924

y =
        -.65084590393626202348291398149803
```

KINEMATICS OF PARTICLES 2

Kinematics involves the study of the motion of bodies irrespective of the forces that may produce that motion. MATLAB can be very useful in solving particle kinematics problems. Problem 2.1 is a rectilinear motion problem illustrating integration with the *int* command. The formulation of this problem results in an equation that cannot be solved exactly except with some rather sophisticated mathematics. When this occurs it is generally easiest to obtain either a graphical or numerical solution. This problem illustrates both approaches. Problem 2.2 is a rectangular coordinates problem that illustrates *int* and *solve*. Problem 2.3 is a relatively straightforward problem where MATLAB is used to generate a plot that might be useful in a parametric study. In problem 2.4, the r-θ components of the velocity are determined using symbolic differentiation (*diff*). The problem also illustrates how computer algebra can simplify what might normally be a rather tedious algebra problem. The *diff* command is further illustrated in problems 2.5 and 2.6. Problem 2.6 is particularly interesting in that it requires differentiation with respect to time of a function whose explicit time dependence is unknown. This happens rather frequently in Dynamics so it is useful to know how to accomplish this with MATLAB.

2.1 Sample Problem 2/4 (Rectilinear Motion)

A freighter is moving at a speed of 8 knots when its engines are suddenly stopped. From this time forward, the deceleration of the ship is proportional to the square of its speed, so that $a = -kv^2$. The sample problem in your text shows that it is rather easy to determine the constant k by measuring the speed of the boat at some specified time. Show how k could be found by (a) measuring the speed after some specified distance and (b) measuring the time required to travel some specified distance. In both cases let the initial speed be v_0.

Problem Formulation

(a) Since time is not involved, the easiest approach is to integrate the equation $vdv = ads$.

$$vdv = ads = -kv^2 ds \qquad \int_{v_0}^{v} \frac{dv}{v} = -k\int_0^s ds \qquad ks = \ln\left(\frac{v_0}{v}\right)$$

With this result it is easy to find k given v at some specified s. To illustrate, assume that $v_0 = 8$ knots and that the speed of the boat is determined to be 3.9 knots after it has traveled one nautical mile.

$$k(1) = \ln\left(\frac{8}{3.9}\right) \qquad k = 0.718 \text{ mi}^{-1}$$

(b) Here we follow the general approach in the sample problem. Integrating $a = dv/dt$ yields

$$\int_{v_0}^{v} \frac{dv}{v^2} = -k\int_0^t dt \qquad -kt = \frac{v - v_0}{vv_0} \qquad v = \frac{v_0}{1 + ktv_0}$$

To obtain the distance s as a function of time we integrate $v = ds/dt$

$$\int_0^s ds = s = \int_0^t vdt = \int_0^t \frac{v_0}{1 + ktv_0} dt \qquad s = \frac{1}{k}\ln(1 + ktv_0)$$

This equation turns out to be very difficult to solve for k. A good mathematician or someone familiar with symbolic algebra software might be able to find the general solution for k in terms of the so-called LambertW function (LambertW(x) is the solution of the equation $ye^y = x$). Even if this solution were found it would be of little use in most practical situations. For example, you would have to spend some time familiarizing yourself with the function. Once this is done you would still have to use a program like Maple or a mathematical handbook to evaluate the function.

For these reasons it is probably easiest to find k either graphically or numerically. Obtaining a numerical solution with MATLAB is so easy that there is little reason not to use this approach. It is generally advisable though to use a graphical approach even when a numerical solution is being obtained. This is the best way to identify whether there are multiple solutions to the problem and also serves as a useful check on the numerical results. Thus, both approaches are illustrated below.

The usual way to generate a graphical solution is to rearrange the equation so as to give a function that is zero at points that are solutions to the original equation. Rearranging the equation above in this manner yields,

$$f = ks - \ln(1 + ktv_0) = 0$$

Given values of s, t, and v_0, f can be plotted versus k. The value of k at which $f = 0$ provides the solution to the original equation.

MATLAB Worksheet and Scripts

Although the integrations are simple in this problem, we'll go ahead and evaluate them symbolically for purposes of illustration.

```
EDU» s_a='-1/k'*int('1/x','x','v0','v')          % part (a)

s_a = -1/k*(log(v)-log(v0))

EDU» s_b=int('v0/(1+k*x*v0)','x',0,'t')     % part (b)

s_b = log(1+t*k*v0)/k
```

The following script implements a graphical solution for part (b). To illustrate, we take v0 = 8 knots and assume that the boat is found to move 1.1 nautical miles after 10 minutes.

```
%%%%%%%%%%% Script %%%%%%%%%%%%%%%%%%%
% This script plots f as a function of k for
% given values of v0, s, and t
v0 = 8;           % initial speed (knots)
s = 1.1;          % distance (nautical miles)
t = 10/60;        % time (hours)
k = 0:0.01:0.5;
f = s*k-log(1+k*v0*t);
line = k*0;
plot(k,f,k,line)
xlabel('k')
ylabel('f')
```

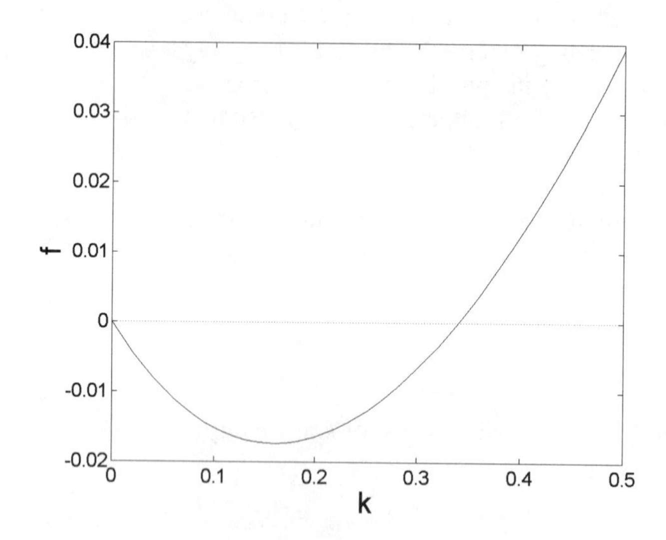

The above graph shows that k is about 0.34 mi^{-1}. Now let's try to find a solution with the symbolic *solve*. First we write a Maple expression for f given our assumed values for v0, s, and t.

EDU» f = '1.1*k - log(1+k*8*10/60) = 0'

f = 1.1*k - log(1+k*8*10/60)=0

EDU» solve(f, 'k')
ans =
[0]
[.33923053342470867736031513009887]

Thus, $k = 0.3392$ mi^{-1}.

2.2 Problem 2/94 (Rectangular Coordinates)

A projectile is fired into an experimental fluid at time t = 0. The initial speed is v_o and the angle to the horizontal is θ. The drag on the projectile results in an acceleration term $\mathbf{a}_D = -k\mathbf{v}$, where k is a constant and **v** is the velocity of the projectile. Determine the x- and y-components of both the velocity and the displacement as functions of time. What is the terminal velocity? For the case $v_o = 40$ m/s and $\theta = 60°$, plot (on a single graph) the trajectory (y versus x) of the projectile for $k = 0.2$, 0.4 and 1 sec^{-1}. Include the effects of gravitational acceleration.

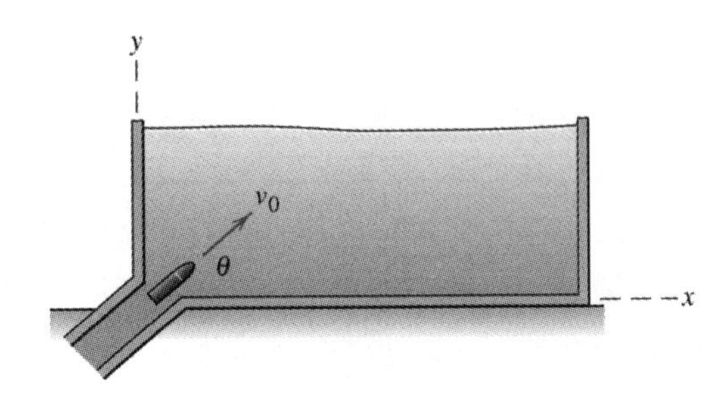

Problem Formulation

The x and y components of the acceleration are:

$$a_x = -k\,v_x \quad \text{and} \quad a_y = -g - k\,v_y$$

These equations can be integrated to yield the velocities. To illustrate, consider

$$a_y = \frac{dv_y}{dt} = -g - k\,v_y\,.$$

Rearranging terms yields $dt = -\dfrac{dv_y}{g + k\,v_y}$. The right hand side is to be integrated from $v_{y0} = v0\,\sin(\theta)$ to v_y, thus

$$t = -\int_{v0\sin\theta}^{v_y} \frac{ds}{g + ks}$$

MATLAB will be used to solve this equation for v_y as a function of time,

$$v_y = \left(v_0 \sin\theta + \frac{g}{k}\right)e^{-kt} - \frac{g}{k}$$

v_y will then be integrated over time to yield y.

$$y = \frac{1}{k}\left(v_0 \sin\theta + \frac{g}{k}\right)\left(1 - e^{-kt}\right) - \frac{g}{k}t$$

An analogous approach is used for the x components yielding,

$$v_x = (v_0 \cos\theta)e^{-kt} \qquad x = \frac{v_0 \cos\theta}{k}\left(1 - e^{-kt}\right)$$

Before proceeding, let's observe that the terminal velocity, which by definition is constant, can be determined by setting the acceleration equal to zero. Thus, the components of the terminal velocity are $v_x = 0$ and $v_y = -\frac{g}{k}$. In other words, at long times the projectile will be moving down at a constant velocity of $\frac{g}{k}$.

MATLAB Scripts

```
%%%%%%%%%%%%% Script #1 %%%%%%%%%%%%%%%%%%%%%%
% This script performs the symbolic integrations
% to obtain x and y as functions of time t.
syms g theta s k t vy vx v0
%%%%%%%%%%%%% y direction %%%%%%%%%%%%%%%%%%%%%
% Note that we re-write the equations in the form
% expression = 0 and then omit the "=0"
eqn1 = t+int(1/(g+k*s),s,v0*sin(theta),vy);
vy = solve(eqn1,vy)
y = int(subs(vy,t,s),s,0,t)
% Note the substitution back to the dummy integration
% variable s.
%
%%%%%%%%%%%% x direction %%%%%%%%%%%%%%%%%%%%%%
eqn2 = t+int(1/(k*s),s,v0*cos(theta),vx);
vx = solve(eqn2,vx)
x = int(subs(vx,t,s),s,0,t)
%%%%%%%%%%%% end of script %%%%%%%%%%%%%%%%%%%
```

Output from script #1

vy =
(exp(-t*k)*g+exp(-t*k)*v0*sin(theta)*k-g)/k

y =
-(exp(-t*k)*g+exp(-t*k)*v0*sin(theta)*k+g*t*k-g-v0*sin(theta)*k)/k^2

vx =
exp(-t*k)*v0*cos(theta)

x =
-v0*cos(theta)*(exp(-t*k)-1)/k

The results for x and y are used in the following script.

```
%%%%%%%%%%%%%% Script #2 %%%%%%%%%%%%%%%%%%
% This script plots y versus x for the specified
% values of k
y=inline('1/k*(40*sin(60*pi/180)+9.81/k)*(1-exp(-k*t))-9.81*t/k');
x=inline('1/k*40*cos(60*pi/180)*(1-exp(-k*t))');
t=0:0.05:10;
plot(x(.2,t),y(.2,t),x(.4,t),y(.4,t),x(1,t),y(1,t))
axis([0 80 0 50])
xlabel('x (m)')
ylabel('y (m)')
%%%%%%%%%%%% end of script %%%%%%%%%%%%%%%%
```

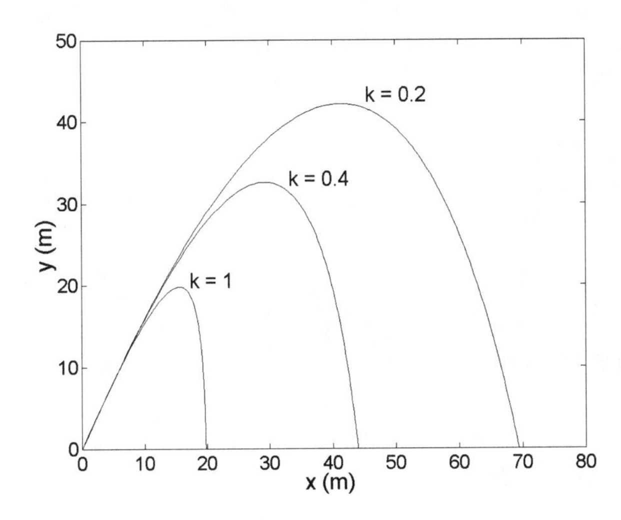

2.3 Problem 2/120 (n-t Coordinates)

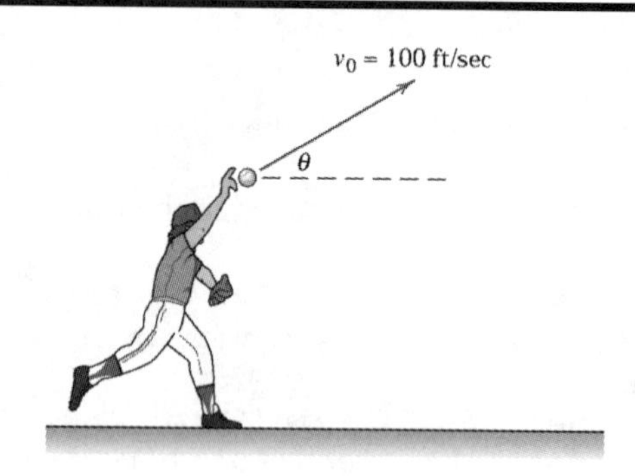

A baseball player releases a ball with initial conditions shown in the figure. Plot the radius of curvature of the path just after release and at the apex as a function of the release angle θ. Explain the trends in both results as θ approaches 90°.

Problem Formulation

Just after release

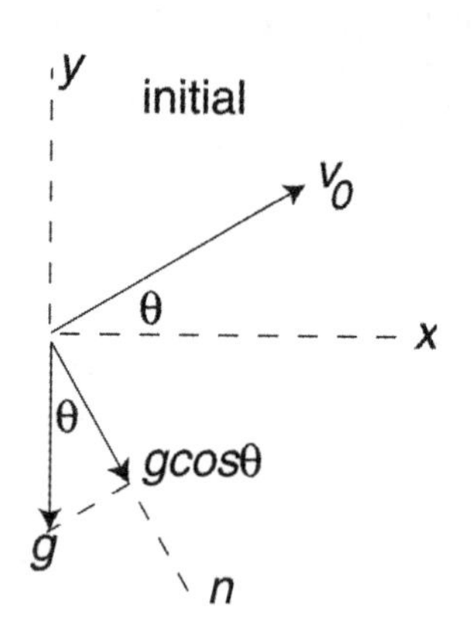

$$a_n = g\cos\theta = \frac{v_0^2}{\rho} \qquad \rho = \frac{v_0^2}{g\cos\theta}$$

At the apex

At the apex, $v_y = 0$ and $v = v_x = v_0\cos\theta$. Since v is horizontal, the normal direction is vertically downward so that $a_n = g$.

$$a_n = g = \frac{(v_0\cos\theta)^2}{\rho} \qquad \rho = \frac{(v_0\cos\theta)^2}{g}$$

MATLAB Script

```
%%%%%%%%%%%%%%%% Script %%%%%%%%%%%%%%%%%%%
v0 = 100;
g = 32.2;
theta = 0:0.01:pi/2;
rho_i = v0^2/g./cos(theta);
rho_a = (v0*cos(theta)).^2/g;
plot(theta*180/pi,rho_i, theta*180/pi,rho_a)
xlabel('theta (deg)')
title('radius of curvature (ft)')
axis([0 90 0 800])
% note that we need to limit the vertical axis since the
% initial rho approaches infinity as theta approaches
% 90 degrees.
%%%%%%%%%%%%%% end of script %%%%%%%%%%%%%%%
```

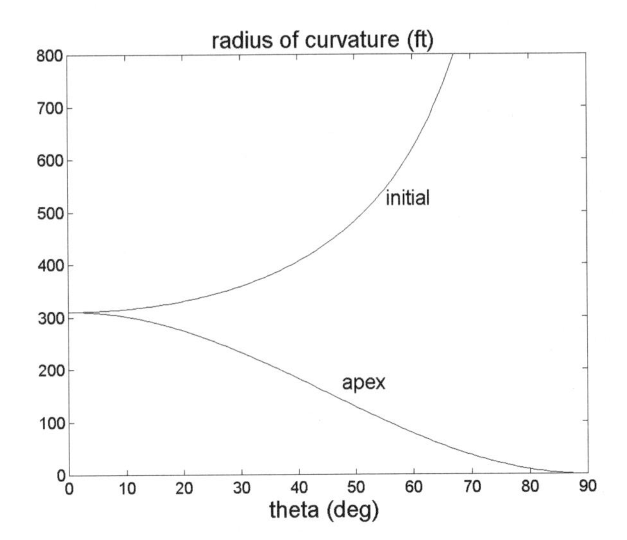

Note that as θ approaches 90°, the initial ρ goes to infinity while ρ at the apex approaches zero. When $\theta = 90°$, the ball travels along a straight (vertical) path. As you recall, straight paths have a radius of curvature of infinity. At the apex, the velocity will be zero giving a radius of curvature of zero.

2.4 Sample Problem 2/10 (Polar Coordinates)

A tracking radar lies in the vertical plane of the path of a rocket which is coasting in unpowered flight above the atmosphere. For the instant when $\theta = 30°$, the tracking data give $r = 25(10^4)$ feet, $\dot{r} = 4000$ ft/s, and $\dot{\theta} = 0.8$ deg/s. Let this instant define the initial conditions at time t = 0 and plot v_r and v_θ as a function of time for the next 150 seconds. You may assume that g remains constant at 31.4 ft/s^2 during this time interval.

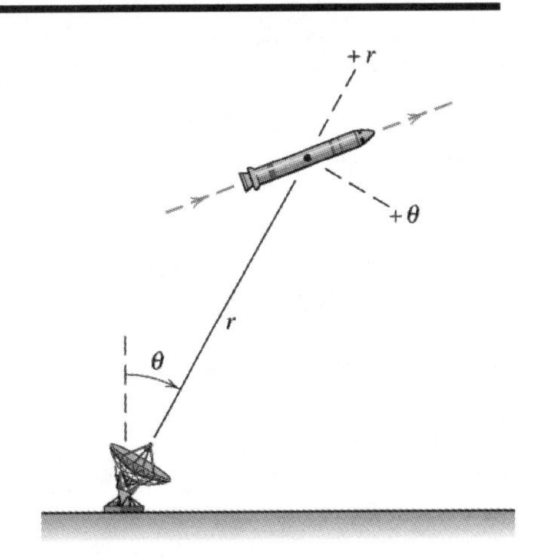

Problem Formulation

Place a Cartesian coordinate system at the radar with x positive to the right and y positive up. Since the rocket is coasting in unpowered flight we can use the equations for projectile motion.

$$x = x_0 + v_0 \cos(\beta)t \qquad y = y_0 + v_0 \sin(\beta)t - \frac{1}{2}gt^2$$

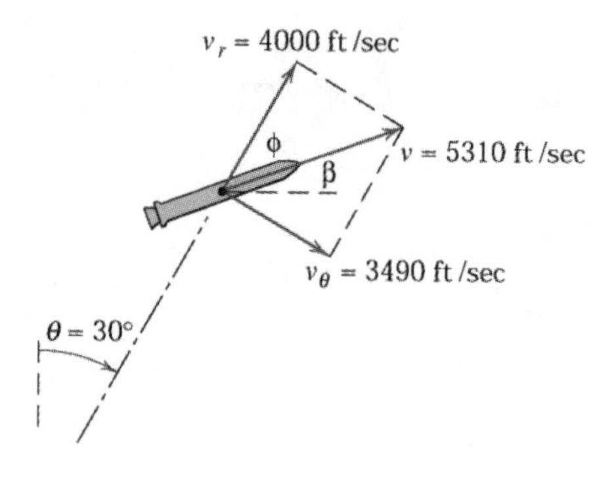

Where $x_0 = 25(10^4)\sin(30)$ ft, $y_0 = 25(10^4)\cos(30)$ ft, v_0 is the initial speed (5310 ft/sec, see the sample problem) and β is the angle that v_0 makes with the horizontal. From the figure shown to the right we can find the angle between v_0 and the r axis as $\phi = \tan^{-1}(3490/4000) = 41.11^\circ$. Since the r axis is 60° from the horizontal, $\beta = 60 - 41.11 = 18.89°$.

With r and θ defined as in the sample problem we have, at any time t

$$r = \sqrt{x^2 + y^2} \qquad \theta = \tan^{-1}(x/y)$$

Now we find v_r and v_θ from their definitions.

$$v_r = \dot{r} \qquad v_\theta = r\dot{\theta}$$

Substitution of x and y into the above equations and carrying out the derivatives with respect to time gives v_r and v_θ as functions of time. The results are very messy and will not be given here. Remember, though, that substitutions such as this can be made automatically when using computer software such as MATLAB.

MATLAB Scripts

```
%%%%%%%%%%%%%%%%%%%%%% Script # 1 %%%%%%%%%%%%%%%%%%%%%%%%%%
% This script obtains symbolic results for the r and
% theta components of the velocity
syms x y t x0 y0 v0 beta g
x = x0+v0*cos(beta)*t
y = y0+v0*sin(beta)*t-1/2*g*t^2
r = sqrt(x^2+y^2)
theta = atan(x/y)
vr = diff(r,t);
vtheta = r*diff(theta,t);
% The results from these operations will be copied and
% pasted into script #2 for plotting. Before they can be
% used they have to be "vectorized", which places periods
% in front of the operators to insure term by term rather
% than matrix operations. We'll go ahead and do that here
% and then copy and paste the vectorized results.
vr = vectorize(vr)
vtheta = vectorize(vtheta)
```

Output from Script #1

x = x0+v0*cos(beta)*t

y = y0+v0*sin(beta)*t-1/2*g*t^2

r = ((x0+v0*cos(beta)*t)^2+(y0+v0*sin(beta)*t-1/2*g*t^2)^2)^(1/2)

theta = atan((x0+v0*cos(beta)*t)/(y0+v0*sin(beta)*t-1/2*g*t^2))

vr =
1./2./((x0+v0.*cos(beta).*t).^2+(y0+v0.*sin(beta).*t-
1./2.*g.*t.^2).^2).^(1./2).*(2.*(x0+v0.*cos(beta).*t).*v0.*cos(beta)+2.*(y0+v0.*
sin(beta).*t-1./2.*g.*t.^2).*(v0.*sin(beta)-g.*t))

vtheta =
((x0+v0.*cos(beta).*t).^2+(y0+v0.*sin(beta).*t-
1./2.*g.*t.^2).^2).^(1./2).*(v0.*cos(beta)./(y0+v0.*sin(beta).*t-1./2.*g.*t.^2)-
(x0+v0.*cos(beta).*t)./(y0+v0.*sin(beta).*t-1./2.*g.*t.^2).^2.*(v0.*sin(beta)-
g.*t))./(1+(x0+v0.*cos(beta).*t).^2./(y0+v0.*sin(beta).*t-1./2.*g.*t.^2).^2)

```
%%%%%%%%%%%%%%%%%%% Script # 2 %%%%%%%%%%%%%%%%%%%
% This script plots the r and theta components of the
% velocity as functions of time.
x0 = 25*10^4*sin(30*pi/180);
y0 = 25*10^4*cos(30*pi/180);
v0 = 5310;
beta = 18.89*pi/180;
g = 31.4;
t = 0:0.1:150;
% The following expressions are copied and pasted from the
% results obtained with script #1.
vr = 1./2./((x0+v0.*cos(beta).*t).^2+(y0+v0.*sin(beta).*t-
1./2.*g.*t.^2).^2).^(1./2).*(2.*(x0+v0.*cos(beta).*t).*v0.*
cos(beta)+2.*(y0+v0.*sin(beta).*t-1./2.*g.*t.^2).*
(v0.*sin(beta)-g.*t));
vtheta = ((x0+v0.*cos(beta).*t).^2+(y0+v0.*sin(beta).*t-
1./2.*g.*t.^2).^2).^(1./2).*(v0.*cos(beta)./(y0+v0.*sin(bet
a).*t-1./2.*g.*t.^2)-(x0+v0.*cos(beta).*t)./
(y0+v0.*sin(beta).*t-1./2.*g.*t.^2).^2.*(v0.*sin(beta)-
g.*t))./(1+(x0+v0.*cos(beta).*t).^2./(y0+v0.*sin(beta).*t-
1./2.*g.*t.^2).^2);
plot(t, vr, t, vtheta)
xlabel('t (sec)')
ylabel('velocity (ft/sec)')
```

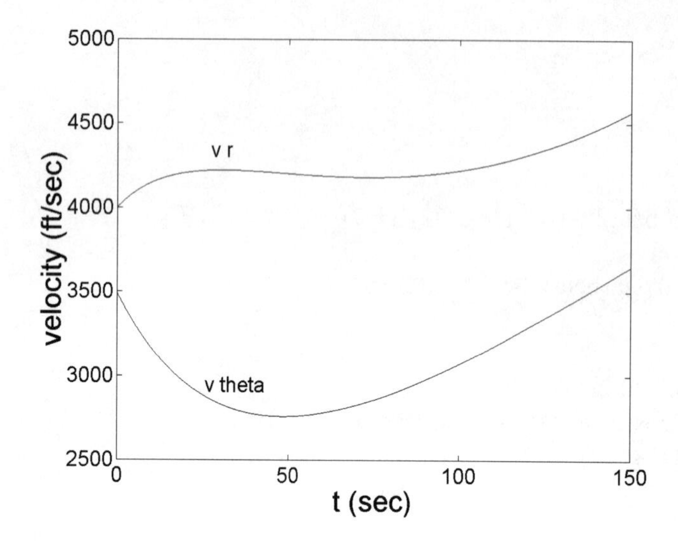

5000
4500
4000
3500
3000
2500
velocity (ft/sec)
v r
v theta
0
50
100
150
t (sec)

2.5 Problem 2/180 (Space Curvilinear Motion)

The base structure of the firetruck ladder rotates about a vertical axis through O with a constant angular velocity $\dot{\theta} = \Omega$. At the same time, the ladder unit OB elevates at a constant rate $\dot{\phi} = \Psi$, and section AB of the ladder extends from within section OA at the constant rate $\dot{R} = \Lambda$. Find general expressions for the components of acceleration of point B in spherical coordinates if, at time $t = 0$, $\theta = 0$, $\phi = 0$, and $AB = 0$. Express your answers in terms of Ω, Ψ, Λ, R_0 and t, where $R_0 = OA$ and is constant. Plot the components of acceleration of B as a function of time for the case Ω=10 deg/s, Ψ= 7 deg/s, Λ = 0.5 m/s, and R_0 = 9 m. Let t vary between 0 and the time at which ϕ = 90°.

Problem Formulation

The components of acceleration in spherical coordinates are,

$$a_R = \ddot{R} - R\dot{\phi}^2 - R\dot{\theta}^2 \cos^2 \phi$$

$$a_\theta = \frac{\cos\phi}{R}\frac{d}{dt}\left(R^2\dot{\theta}\right) - 2R\dot{\theta}\dot{\phi}\sin\phi$$

$$a_\phi = \frac{1}{R}\frac{d}{dt}\left(R^2\dot{\phi}\right) + R\dot{\theta}^2 \sin\phi\cos\phi$$

The components may be obtained as functions of time by substituting,

$$R = R_0 + \Lambda t\,,\ \theta = \Omega t \ \text{ and } \ \phi = \Psi t$$

Differentiation and substitution will be performed in MATLAB. The results are,

$$a_R = (R_0 + \Lambda t)\left(\Psi^2 - \Omega^2 \cos^2(\Psi t)\right)$$

$$a_\theta = 2\Omega\Lambda\cos(\Psi t) - 2\Omega\Psi(R_0 + \Lambda t)\sin(\Psi t)$$

$$a_\phi = 2\Psi\Lambda + (R_0 + \Lambda t)\Omega^2 \sin(\Psi t)\cos(\Psi t)$$

MATLAB Scripts

```
%%%%%%%%%%%%%%%%%%%%%%% Script #1 %%%%%%%%%%%%%%%%%%%%%%%%%%%%
% This script calculates the components of the
% acceleration symbolically
% O = Omega; P = Phi; L = Lambda
syms O P L t R0
R = R0+L*t;
theta = O*t;
phi = P*t;
% Even though there are some obvious simplifications
% in this case, we still write the most general
% expressions for the spherical components of the
% acceleration. In this way we can consider other types
% of time dependence without modifying the script.
a_R = diff(R,t,2)-R*diff(phi,t)^2-R*diff(theta,t)^2
*cos(phi)^2
a_theta = cos(phi)/R*diff(R^2*diff(theta,t),t)-
2*R*diff(theta,t)*diff(phi,t)*sin(phi)
a_phi = 1/R*diff(R^2*diff(phi,t),t)+R*diff(theta,t)^2
*sin(phi)*cos(phi)
%%%%%%%%%%%%%%%%%% end of script %%%%%%%%%%%%%%%%%%%%%%%%%%%%
```

Output of script #1

a_R =
-(R0+L*t)*P^2-(R0+L*t)*O^2*cos(P*t)^2

a_theta =
2*cos(P*t)*O*L-(2*R0+2*L*t)*O*P*sin(P*t)

a_phi =
2*P*L+(R0+L*t)*O^2*sin(P*t)*cos(P*t)

```
%%%%%%%%%%%%%%%%%%%%%%% Script #2 %%%%%%%%%%%%%%%%%%%%%%%%%%%%
% This script plots the components of the acceleration
% as functions of time
% O = Omega; P = Phi; L = Lambda
L = 0.5; O = 10*pi/180;
P = 7*pi/180; R0 = 9;
tf = pi/2/P; % time at which phi=pi/2
t=0:0.05:tf;
a_R = -(R0+L*t)*P^2-(R0+L*t).*O^2.*cos(P*t).^2;
a_theta = 2*cos(P*t)*O*L-(2*R0+2*L*t)*O*P.*sin(P*t);
a_phi = 2*P*L+(R0+L*t)*O^2.*sin(P*t).*cos(P*t);
plot(t,a_R,t,a_theta,t,a_phi)
```

```
xlabel('time (sec)')
title('acceleration (m/s^2)')
%%%%%%%%%%%%%%% end of script %%%%%%%%%%%%%%%%%%%%%%%%%%%%%
```

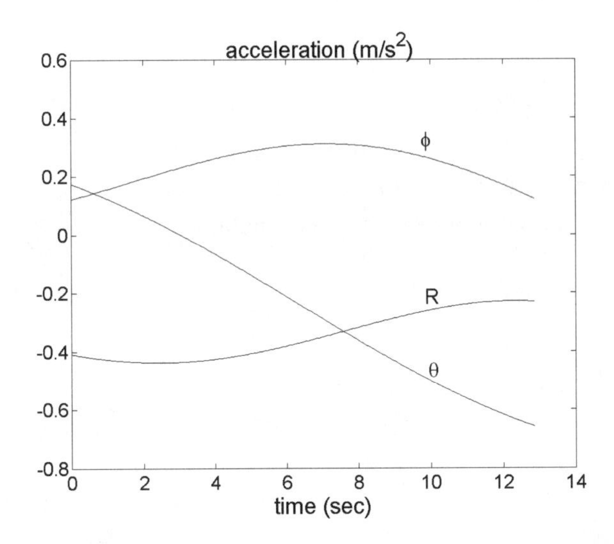

2.6 Sample Problem 2/15 (Constrained Motion of Connected Particles)

The tractor A is used to hoist the bale B with the pulley arrangement shown. If A has a forward velocity v_A, determine an expression for the upward velocity v_B of the bale in terms of x. Put the result in nondimensional form by introducing the velocity ratio $\eta = v_B/v_A$ and nondimensional position $\chi = x/h$. Plot η versus χ for $0 \le \chi \le 2$.

Problem Formulation

The length L of the cable can be written

$$L = 2(h - y) + l + constants = 2(h - y) + \sqrt{h^2 + x^2} + constants$$

Now, $\dot{L} = 0$ will be used to obtain a relation between v_A ($= \dot{x}$) and v_B ($= \dot{y}$).

$$\dot{L} = 0 = -2\dot{y} + \frac{x\dot{x}}{\sqrt{h^2 + x^2}} \qquad v_B = \frac{1}{2}\frac{x v_A}{\sqrt{h^2 + x^2}}$$

The nondimensional result is now obtained by substituting $v_B = \eta v_A$ and $x = \chi h$.

$$\eta = \frac{1}{2}\frac{\chi}{\sqrt{1+\chi^2}}$$

Even though these operations are rather easily performed by hand, it is instructive to have MATLAB do them. In particular, it will be instructive to see how to evaluate $\dot{L}$ even though x and y are not known explicitly as functions of time.

MATLAB Worksheet and Script

```
EDU» syms t h vA vB chi eta
EDU» x = sym('x(t)'); y = sym('y(t)');
EDU» L = 2*(h-y)+sqrt(h^2+x^2)
L =
2*h-2*y(t)+(h^2+x(t)^2)^(1/2)
```

Note that we need to differentiate L with respect to time. Both x and y depend on time, however, exactly how they depend on time is not known. It turns out that this is not a problem. All we need to do is let MATLAB know x and y depend on time by writing x = sym('x(t)') and y = sym('y(t)'). To illustrate, consider the following.

```
EDU» diff(x, t)
ans =
diff(x(t),t)
```

This derivative would have been evaluated as zero had we not declared x to be a function of t. Now let's proceed by differentiating L with respect to t. We give the result a name (*eqn*) to facilitate later substitutions.

```
EDU» eqn = diff(L,t)
eqn =
-2*diff(y(t),t)+1/(h^2+x(t)^2)^(1/2)*x(t)*diff(x(t),t)
```

EDU» eqn = subs(eqn,diff(y,t),eta*vA) % substitutes $v_B = \eta v_A$ for $\dot{y}$ (diff(y,t))

```
eqn =
-2*eta*vA+1/(h^2+x(t)^2)^(1/2)*x(t)*diff(x(t),t)
```

```
EDU» eqn = subs(eqn,diff(x,t),vA)
eqn =
```

-2*eta*vA+1/(h^2+x(t)^2)^(1/2)*x(t)*vA

EDU» eqn = subs(eqn,x,chi*h)
eqn =
-2*eta*vA+1/(h^2+chi^2*h^2)^(1/2)*chi*h*vA

Now let's remember that *eqn* is just a name for dL/dt which is zero. We now solve this equation for η (eta). Also recall that solve(eqn, eta) actually solves the equation eqn = 0 for eta.

EDU» eta = solve(eqn, eta)
eta =
1/2*chi*h/(h^2+chi^2*h^2)^(1/2)

We note finally that the h cancels in the above expression yielding the result given in the problem formulation section above. Now we can produce the required plot.

```
%%%%%%% Script for plotting chi versus eta %%%%%
chi = 0:0.01:2;
eta = 1/2*chi./sqrt(1+chi.^2);
plot(chi,eta)
xlabel('x/h')
title('v_B/v_A')
%%%%%%%%%%%%%%%% end of script %%%%%%%%%%%%%%%%
```

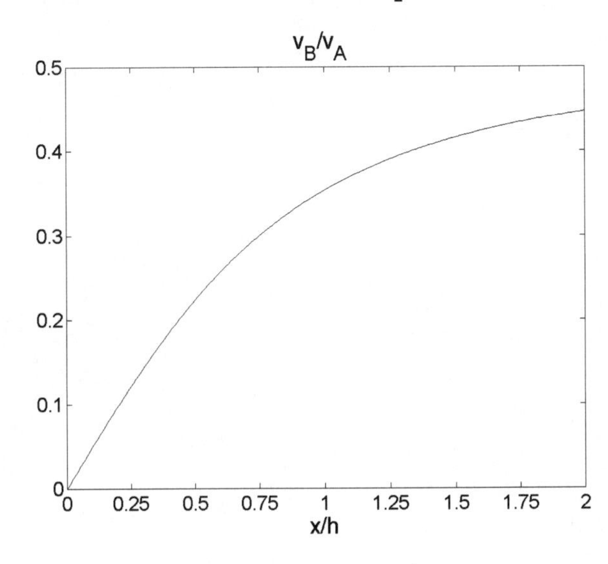

KINETICS OF PARTICLES

3

The kinetics of particles is concerned with the motion produced by unbalanced forces acting on a particle. This chapter considers three approaches to the solution of particle kinetics problems: (1) direct application of Newton's second law, (2) work and energy, and (3) impulse and momentum. Problem 3.1 is a rectilinear motion problem where *solve* is used solve three equations symbolically for three unknowns. In problem 3.2, *dsolve* is used to solve a second order differential equation with initial conditions. The absolute path of a particle is then plotted using *polar*. Problem 3.3 uses MATLAB to study the effect of initial spring stretch upon the velocity of a slider. A physical interpretation of the results is also required. Problem 3.4 is a typical ballistic pendulum problem requiring both work/energy and conservation of momentum to relate the velocity of a projectile to the maximum swing angle of a pendulum. Problem 3.5 is a relatively straightforward conservation of angular momentum problem where MATLAB is used to generate a plot that might be useful in a parametric study. In problem 3.6, two equations are solved symbolically for two unknowns using *solve*. The maximum value of a function is then determined using *diff* and *solve*.

3.1 Sample Problem 3/3 (Rectilinear Motion)

The 250-lb concrete block A is released from rest in the position shown and pulls the 400-lb log up the 30° ramp. Plot the velocity of the block as it hits the ground at B as a function of the coefficient of kinetic friction μ_k between the log and the ramp. Let μ_k vary between 0 and 1. Why does the computer not plot results for the entire range specified?

Problem Formulation

The constant length of the cable is $L = 2s_C + s_A$ (see figure). Differentiating this expression twice yields a relation between the acceleration of A and C (note that $a_C = a_{LOG}$).

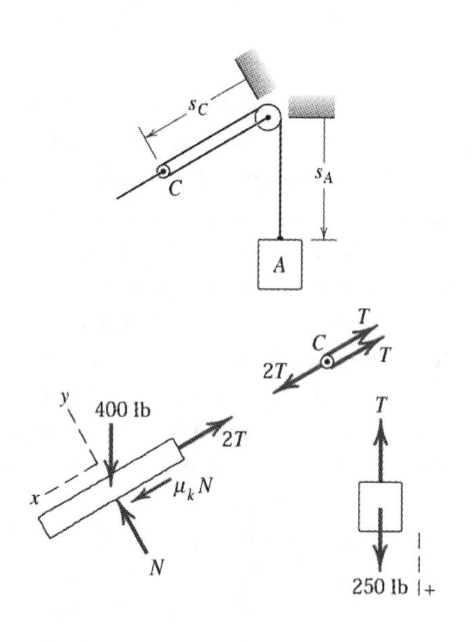

$$0 = 2a_C + a_A \qquad (1)$$

From the free-body diagram for the log

$$[\Sigma F_y = ma_y = 0] \quad N - 400\cos(30) = 0$$

$$[\Sigma F_x = ma_x] \qquad \mu_k N - 2T + 400\sin(30) = \frac{400}{32.2} a_C$$

Substituting N yields,

$$400\mu_k \cos(30) - 2T + 400\sin(30) = \frac{400}{32.2} a_C \qquad (2)$$

From the free-body diagram for block A

$$[\downarrow \Sigma F = ma] \qquad 250 - T = \frac{250}{32.2} a_A \qquad (3)$$

MATLAB will be used to solve the three equations above for a_A, a_C and T in terms of μ_k. Since the accelerations are constant, $v_A^2 = 2a_A d$ where d is the vertical distance through which block A has fallen. Thus, the velocity of A when it strikes the ground ($d = 20$ ft) is

$$v_{Af} = \sqrt{40a_A} = v_B$$

MATLAB Scripts

```
%%%%%%%%%%%%%%%% Script #1 %%%%%%%%%%%%%%%%%%%%%%%%
% This script solves the three equations developed
% in the problem formulation section for the two
% accelerations (aC and aA) and the tension T.
% When solving multiple equations, it is a good idea
% to let the variables solved for be single characters.
% Thus, we set x = aC, y = aA, z = T.
eqn1 = '2*x+y = 0';
eqn2 = '400*muk*cos(theta)-2*z+400*sin(theta)=400/32.2*x';
eqn3 = '250-z = 250/32.2*y';
[x,y,z]=solve(eqn1,eqn2,eqn3)
%%%%%%%%%%%%% end of script %%%%%%%%%%%%%%%%%%%%%%%%
```

Output from Script #1

x = 7.9674*muk - 6.9

y = -15.9349*muk + 13.8

z = 142.857+123.7179*muk

The only thing we are interested in here is a_A (y in our solution above). Thus,

$$a_A = 13.8 - 15.9349\mu_k$$

Note that the accelerations may be either positive or negative depending on the value of μ_k. The largest value of μ_k for which the block will move up can thus be found by solving the equation $a_A = 0$ for μ_k. This yields μ_k = 13.8/15.935 = 0.866.

```
%%%%%%%%%%%%%%%%%%% Script #2 %%%%%%%%%%%%%%%%%%%%%%%%%%
% This script plots the vB (the velocity of the
% block as it hits the ground) versus the friction
% coeficient mu_k
muk = 0:0.01:1;
aA=-15.9349*muk+13.8;
vB = sqrt(40*aA)
plot(muk,vB)
xlabel('mu_k')
ylabel('v_B (ft/sec)')
%%%%%%%%%%%%%% end of script %%%%%%%%%%%%%%%%%%%%%%%%%%%
```

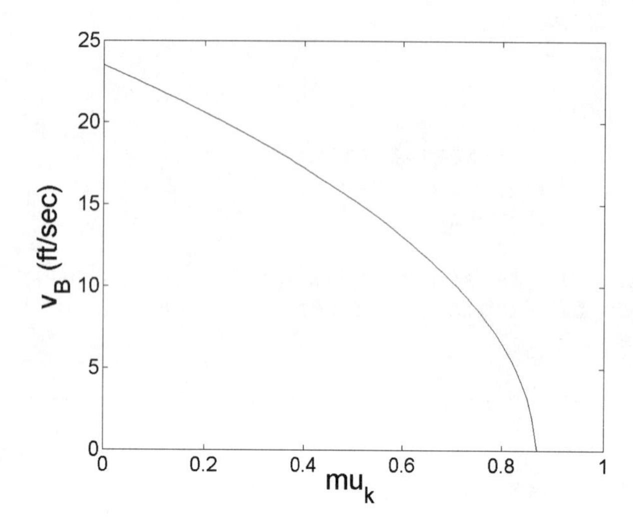

At first sight, it seems that no results are plotted beyond the limiting value for μ_k (0.866) that was determined above. If the plot were in color, you would see that MATLAB has actually plotted zeros beyond this point. From a numerical point of view this occurs because MATLAB will plot only the real part of complex numbers. If you have MATLAB print the values of vB you will find that the real parts of all the complex numbers generated are zero. Whenever imaginary or complex values result there is usually some physical explanation. In this problem, the physical explanation is that the log will not slide up the incline if the coefficient of friction is too large.

3.2 Problem 3/92 (Curvilinear Motion)

The particle P is released at time $t = 0$ from the position $r = r_0$ inside the smooth tube with no velocity relative to the tube, which is driven at the constant angular velocity ω_0 about the vertical axis. Determine the radial velocity v_r, the radial position r, and the transverse velocity v_θ as functions of time t. Plot the absolute path of the particle during the time that it is inside the tube for $r_0 = 0.1$ m, $l = 1$ m, and $\omega_0 = 1$ rad/s.

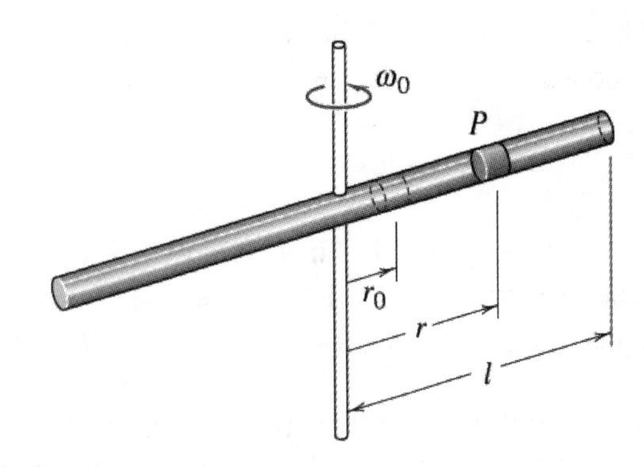

Problem Formulation

From the free-body diagram to the right,

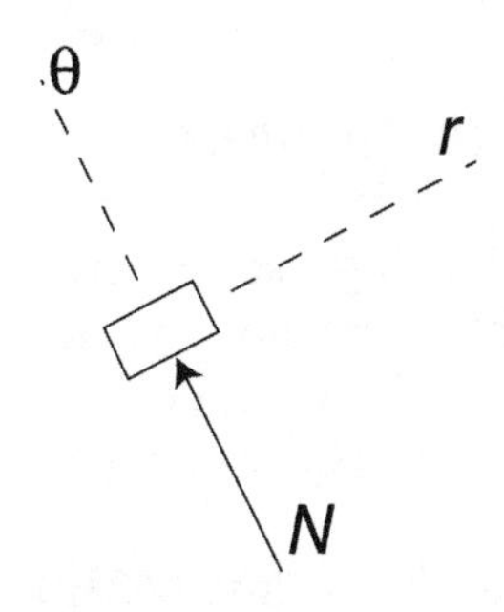

$$\Sigma F_r = 0 = ma_r = m\left(\ddot{r} - r\dot{\theta}^2\right)$$

$$\ddot{r} = r\dot{\theta}^2 = r\omega_0^2$$

Any book on differential equations will have the solution to this equation in terms of the hyperbolic sine and cosine,

$$r = A\sinh(\omega_0 t) + B\cosh(\omega_0 t)$$

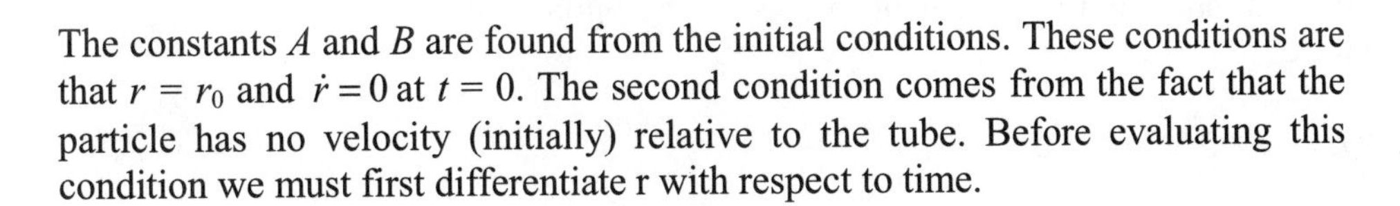

The constants A and B are found from the initial conditions. These conditions are that $r = r_0$ and $\dot{r} = 0$ at $t = 0$. The second condition comes from the fact that the particle has no velocity (initially) relative to the tube. Before evaluating this condition we must first differentiate r with respect to time.

$$\dot{r} = A\omega_0\cosh(\omega_0 t) + B\omega_0\sinh(\omega_0 t)$$

$$r(t=0) = r_0 = A\sinh(0) + B\cosh(0) = B$$
$$\dot{r}(t=0) = 0 = A\omega_0\cosh(0) + B\omega_0\sinh(0) = A\omega_0$$

From the above we have $B = r_0$ and $A = 0$. Thus,

$$r = r_0\cosh(\omega_0 t)$$

From this we can obtain the radial and transverse velocities,

$$v_r = \dot{r} = r_0\omega_0 \sinh(\omega_0 t) \qquad v_\theta = r\dot{\theta} = r_0\omega_0 \cosh(\omega_0 t)$$

The absolute path of the particle will be graphed using polar plotting. For this we need r as a function of θ. Since $\theta = \omega_0 t$ we have,

$$r = r_0 \cosh(\theta)$$

We want to plot this function only up to the point where the particle leaves the tube. Substituting $r = 1$ we have $1 = 0.1\cosh(\theta)$, or $\theta = \cosh^{-1}(10) = 2.993$ rads. Thus, the particle leaves the tube when $\theta = 2.993$ rads (171.5°).

As you will see in the worksheet below, MATLAB can also be used to solve the second order differential equation with initial conditions, greatly simplifying this problem.

MATLAB Worksheet

The following uses *dsolve* to solve the differential equation above. D is the differential operator so that D2r is the second derivative with respect to t (unless specified otherwise, derivatives are assumed to be taken with respect to t). Also note that the expression must be placed in quotation marks.

```
EDU» dsolve('D2r=w0^2*r')
ans =
C1*sinh(w0*t)+C2*cosh(w0*t)
```

This is the general symbolic result found above. C1 and C2 are found from the initial conditions. It is much more convenient to specify the initial conditions in the *dsolve* command.

```
EDU» dsolve('D2r=w0^2*r','r(0)=r0, Dr(0)=0')
ans =
r0*cosh(w0*t)
```

Note how the initial conditions are specified. Again, D is a differential operator so that D(r)(0) = 0 is the initial condition $\dot{r} = 0$ at t =0.

Now we use *polar* to plot the position of the particle.

```
EDU» theta = 0:0.01:2.993;
EDU» r = 0.1*cosh(theta);
EDU» polar(theta,r)
```

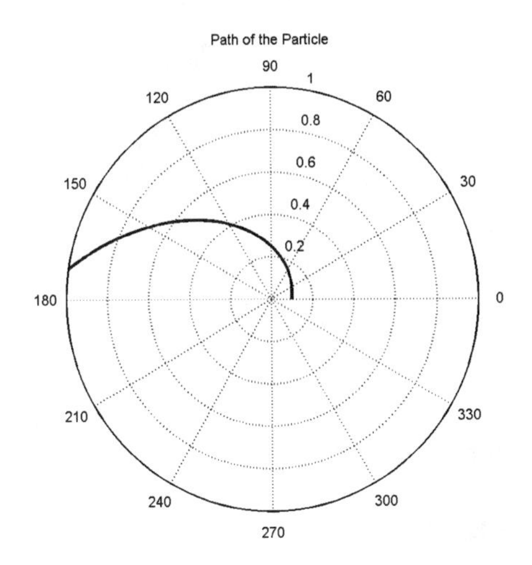

3.3 Sample Problem 3/16 (Potential Energy)

The 10-kg slider A moves with negligible friction up the inclined guide. The attached spring has a stiffness of 60 N/m and is stretched δ m at position A, where the slider is released from rest. The 250-N force is constant and the pulley offers negligible resistance to the motion of the cord. Plot the velocity of the slider as it passes C as a function of the initial spring stretch δ. Let δ vary between –0.4 and 0.8 m and explain the results when δ exceeds a value of about 0.65 m.

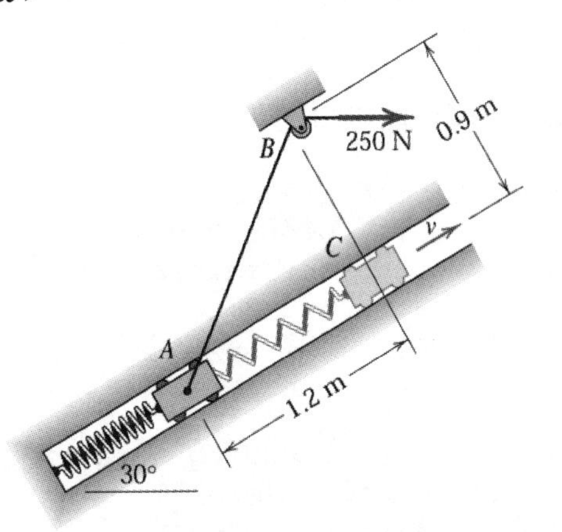

Problem Formulation

The change in the elastic potential energy is

$$\Delta V_e = \frac{1}{2}k\left(x_2^2 - x_1^2\right) = \frac{1}{2}k\left((1.2+\delta)^2 - \delta^2\right)$$

The other results in the sample problem are unchanged,

$$U'_{1-2} = 250(0.6) = 150\ \text{J} \qquad \Delta T = \frac{1}{2}m\left(v^2 - v_0^2\right) = \frac{1}{2}(10)v^2$$

$$\Delta V_g = mg\Delta h = 10(9.81)(1.2\sin 30) = 58.9\ \text{J}$$

$$U'_{1-2} = 150 = \frac{1}{2}(10)v^2 + 58.9 + \frac{1}{2}(60)\left((1.2+\delta)^2 - \delta^2\right)$$

This equation can be solved for v either by hand or by using MATLAB. The result is

$$v = \frac{1}{10}\sqrt{958 - 1440\delta}$$

MATLAB Worksheet and Script

First we will solve the work/energy equation symbolically for v.

EDU» eqn='150=5*v^2+58.9+30*((1.2+del)^2-del^2)'

eqn = 150=5*v^2+58.9+30*((1.2+del)^2-del^2)

EDU» solve(eqn,'v')
ans =
[1/10*(958-1440*del)^(1/2)]
[-1/10*(958-1440*del)^(1/2)]

```
%%%%%%%%%%%%%%%%% Script %%%%%%%%%%%%%%%%%
% This script plots the velocity of the slider
% at C versus the initial spring stretch del
del = -0.4:0.01:0.8;
v = 1/10*sqrt(958-1440*del);
plot(del,v)
axis([-0.4 0.8 0 4])
xlabel('initial spring stretch (m)')
ylabel('velocity (m/s)')
%%%%%%%%%%%%%% end of script %%%%%%%%%%%%%%%
```

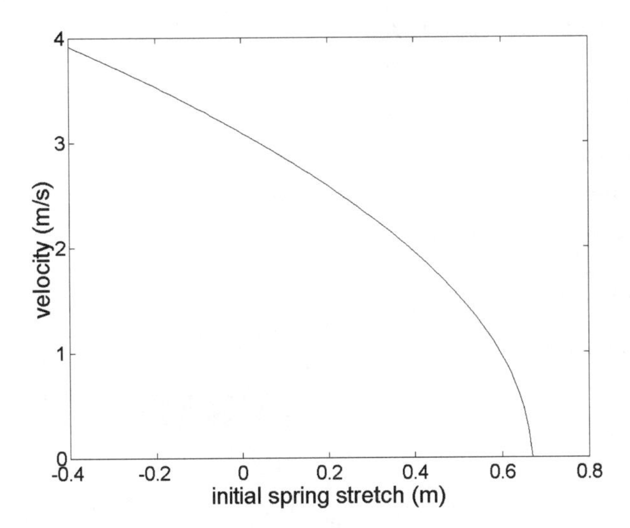

At first sight, it seems that no results are plotted beyond $\delta \cong 0.65$ m. If the plot were in color, you would see that MATLAB has actually plotted zeros beyond this point. Why? One reason is to observe from the above equation that v becomes imaginary when $\delta > 958/1440 = 0.665$ m. MATLAB interprets imaginary numbers as complex numbers with a real part equal to zero. When asked to plot a complex number it will plot only the real part. Thus, we have zeros plotted after $\delta = 0.665$ m. But this is a numerical reason instead of a physical explanation. Usually, imaginary answers signify a situation that is physically impossible for some reason. One way of understanding this is as follows. If the spring is initially compressed it will, at least for some part of the motion, be pushing up and thus be aiding the 250 N force in overcoming the weight of the slider. If the spring is initially stretched, it will always be pulling back on the slider. Thus the 250 N force will have to overcome not only the weight but also the spring force. It stands to reason then that there will be some value for the initial spring stretch beyond which the 250 N force will not be able to pull the slider all the way to C. This value is found from the limiting case where $v = 0$. Thus, the block never reaches C if $\delta > 0.665$ m.

3.4 Problem 3/208 (Linear Impulse/Momentum)

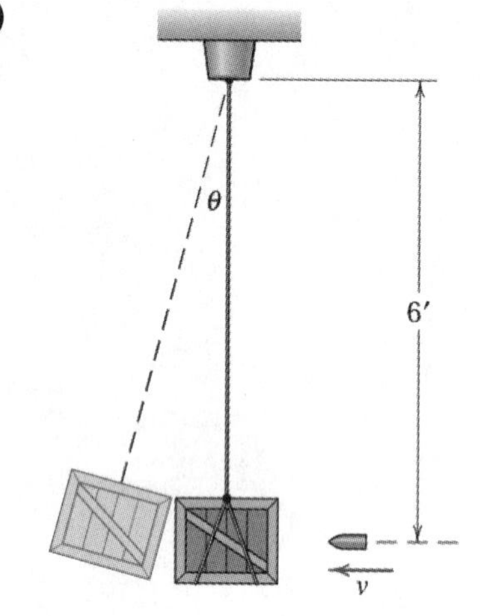

The ballistic pendulum is a simple device to measure the projectile velocity v by observing the maximum angle θ to which the box of sand with embedded particle swings. As an aid for a laboratory technician, make a plot of the velocity v in terms of the maximum angle θ. Assume that the weight of the box is 50-lb while the weight of the projectile is 2-oz.

Problem Formulation

(1) Impulse/Momentum

During impact, $\Delta G = 0$ and $G_1 = G_2$

$$\left(\frac{2/16}{32.2}\right)v + \left(\frac{50}{32.2}\right)(0) = \left(\frac{2/16+50}{32.2}\right)v_b$$

$$v = 401v_b$$

where v is the velocity of the projectile while v_b is the velocity of the box of sand immediately after impact.

(2) Work/Energy

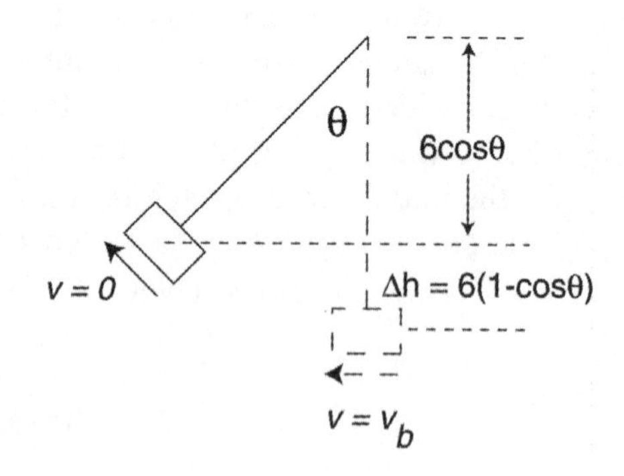

Now we use the work/energy equation with our initial position being the position where the pendulum is still vertical ($\theta = 0$) and the final position is that where the pendulum has rotated through the maximum angle θ.

$$U'_{1-2} = 0 = \Delta T + \Delta V_g = \frac{1}{2}m\left(0^2 - v_b^2\right) + mg\Delta h$$

where m is the combined mass of the box and the projectile.

$$v_b = \sqrt{2g\Delta h} = \sqrt{2(32.2)(6)(1-\cos\theta)}$$

$$v = 401v_b = 7882\sqrt{1-\cos\theta}$$

MATLAB Worksheet

```
EDU» theta = 0:0.01:pi/2;
EDU» v = 7882*sqrt(1-cos(theta));
EDU» plot(theta*180/pi,v)
EDU» xlabel('theta (degrees)')
EDU» title('velocity of projectile (ft/s)')
```

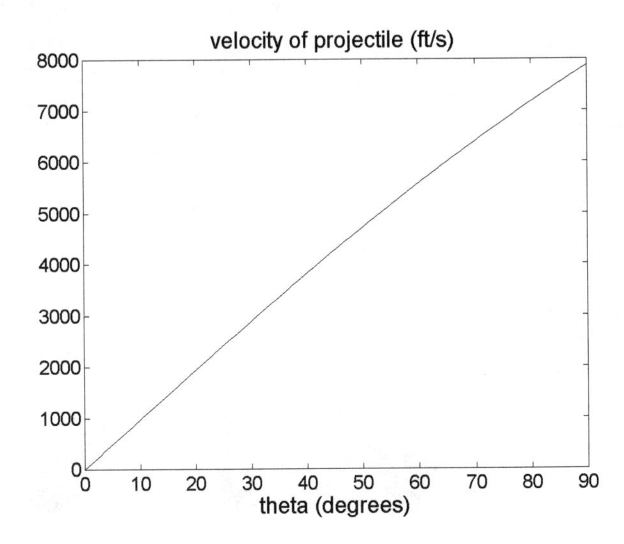

3.5 Problem 3/243 (Angular Impulse/Momentum)

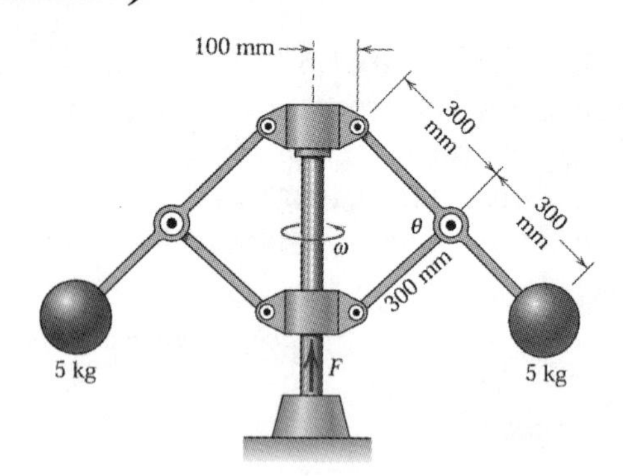

The assembly of two 5-kg spheres is rotating freely about the vertical axis at 40 rev/min with $\theta = 90°$. The force F that maintains the given position is increased to raise the base collar and reduce the angle from 90° to an arbitrary angle θ. Determine the new angular velocity ω and plot ω as a function of θ for $0 \le \theta \le 90°$. Assume that the mass of the arms and collars is negligible.

Problem Formulation

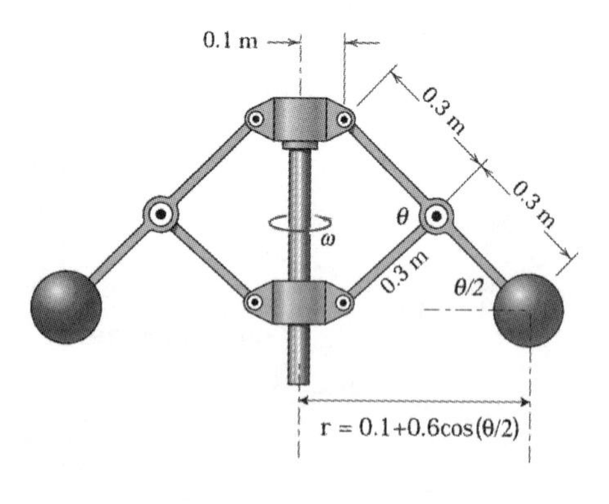

Since the summation of moments about the vertical axis is zero we have conservation of angular momentum about that axis. The spheres are rotating in a circular path about the vertical axis. The angular momentum of a particle moving in a circular path of radius r with angular velocity ω is $H = mr^2\omega$. Thus, from the conservation of angular momentum we have,

$$2mr_0^2\omega_0 = 2mr^2\omega \qquad \omega = \frac{r_0^2}{r^2}\omega_0$$

where $\quad r_0 = 0.1 + 0.6\cos(45^0) \quad$ and $\quad r = 0.1 + 0.6\cos(\theta/2)$

MATLAB Script

```
%%%%%%%%%%% script %%%%%%%%%%%%%%
theta = 0:0.05:pi/2;
r0 = 0.1+0.6*cos(pi/4);
r = 0.1+0.6*cos(theta/2);
w0 = 40*2*pi/60;
w = r0^2./r.^2*w0;
plot(theta*180/pi, w)
xlabel('theta (degrees)')
ylabel('omega (rad/s)')
%%%%%%%% end of script %%%%%%%%%%
```

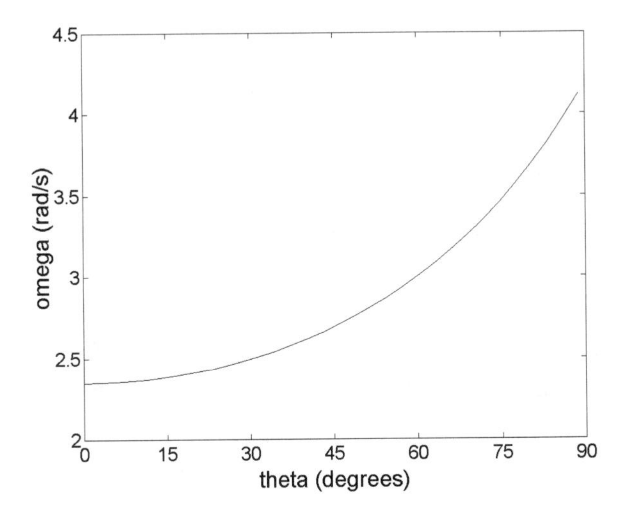

This diagram "begins" at $\theta = 90°$ where $\omega = \omega_0 = 40(2\pi)/60 = 4.19$ rad/s.

3.6 *Problem 3/358 (Curvilinear Motion)*

The 26-in. drum rotates about a horizontal axis with a constant angular velocity $\Omega = 7.5$ rad/sec. The small block A has no motion relative to the drum surface as it passes the bottom position $\theta = 0$. Determine the coefficient of static friction μ_s that would result in block slippage at an angular position θ; plot your expression for $0 \le \theta \le 180°$. Determine the minimum required coefficient value μ_{min} that would allow the block to remain fixed relative to the drum throughout a full revolution. For a friction coefficient slightly less than μ_{min}, at what angular position θ would slippage occur?

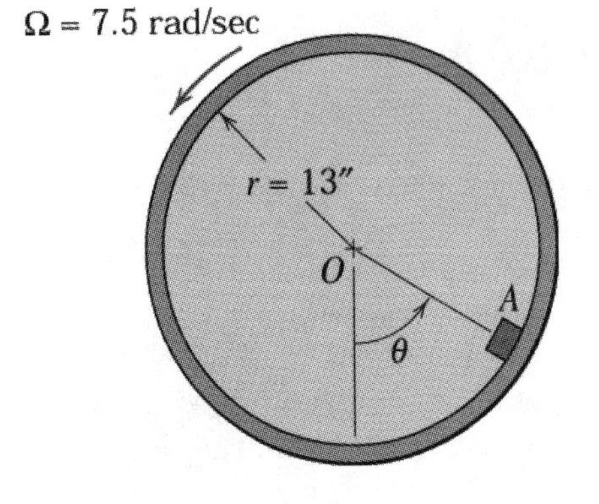

Problem Formulation

From the free body and mass acceleration diagrams,

FBD MAD

$$[\Sigma F_n = ma_n] \qquad N - mg\cos\theta = mr\Omega^2$$

$$[\Sigma F_t = ma_t] \qquad F - mg\sin\theta = 0$$

For impending slip we have $F = \mu_s N$. Substituting F into the above and solving gives,

$$\mu_s = \frac{g \sin\theta}{g\cos\theta + r\Omega^2} = \frac{\sin\theta}{1.8925 + \cos\theta}$$

The last two questions can be answered only after plotting μ_s as a function of θ.

MATLAB Worksheet and Scripts

```
%%%%%%%%%%%%%%% Script #1 %%%%%%%%%%%%%%%%%%%%
% This script solves our two equations symbolically
% for mu_s and N
% O = Omega
% x = mu_s
% y = N
syms theta O g m r x y
eqn1 = y-m*g*cos(theta)-m*r*O^2;
eqn2 = x*y-m*g*sin(theta);
% remember that we write our equations in the form
% expression = 0 and then omit the "=0"
[x,y] = solve(eqn1,eqn2)
%%%%%%%%%%%%%%end of script %%%%%%%%%%%%%%%%%%%%
```

Output of script #1

x =
g*sin(theta)/(g*cos(theta)+r*O^2)

y =
m*g*cos(theta)+m*r*O^2

```
%%%%%%%%%%%%%%% Script #2 %%%%%%%%%%%%%%%%%%%%
% This script plots mu_s as a function of theta
theta = 0:0.02:pi;
mu_s = sin(theta)./(1.8925+cos(theta));
plot(theta*180/pi, mu_s)
xlabel('theta (degrees)')
title('coefficient of static friction')
%%%%%%%%%%%%%%end of script %%%%%%%%%%%%%%%%%%%%
```

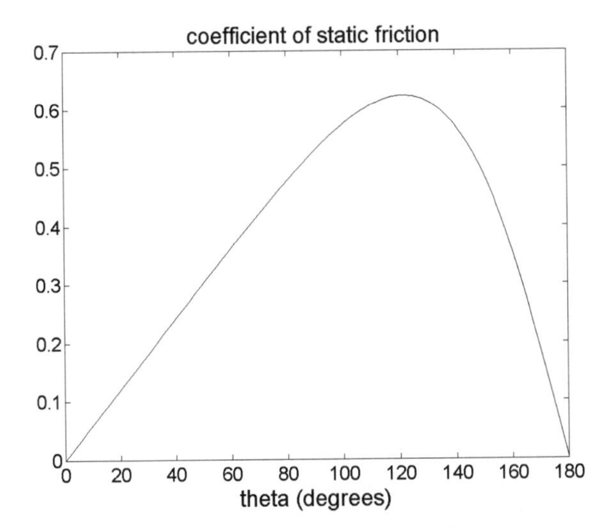

If the block is not to slip at any angle θ, the coefficient of friction must be greater than or equal to any value shown on the plot above. Thus, the minimum required coefficient value μ_{min} that would allow the block to remain fixed relative to the drum throughout a full revolution is equal to the maximum value in the plot above. The location where this maximum occurs can be found by solving the equation $d\mu_s / d\theta = 0$ for θ. This θ can then be substituted into μ_s to yield the required value for μ_{min}. Here's how we do this with MATLAB.

```
EDU» syms theta
EDU» mu_s = sin(theta)./(1.8925+cos(theta));
EDU» dmu = diff(mu_s,theta)
dmu = cos(theta)/(757/400+cos(theta))+sin(theta)^2/(757/400+cos(theta))^2

EDU» theta_m = solve(dmu,theta)
theta_m =
[ -atan(1/302800*236697316401^(1/2))+pi]
[  atan(1/302800*236697316401^(1/2))-pi]

EDU» eval(theta_m)
ans =
   2.1275
  -2.1275

EDU» mu_min = subs(mu_s,theta,2.1275)
mu_min = 0.6224
```

From the above we see that $\mu_{min} = 0.622$. If μ_s is slightly less than this value, the block will slip when $\theta = 2.128$ rads (121.9°).

KINETICS OF SYSTEMS OF PARTICLES

4

This chapter concerns the extension of principles covered in chapters two and three to the study of the motion of general systems of particles. The chapter first considers the three approaches introduced in chapter 3 (direct application of Newton's second law, work/energy, and impulse/momentum) and then moves to other applications such as steady mass flow and variable mass. Problem 4.1 considers an application of the conservation of momentum to a system comprised of a small car and an attached rotating sphere. MATLAB is used to plot the velocity of the car as a function of the angular position of the sphere. The absolute position of the sphere is also plotted. Problem 4.2 uses the concept of steady mass flow to study the effects of geometry upon the design of a sprinkler system. One of the main purposes of this problem is to illustrate how a problem can be greatly simplified using non-dimensional analysis. In particular, an equation containing seven parameters is reduced to a non-dimensional equation with only three parameters. Problem 4.3 is a variable mass problem in which MATLAB is used to integrate the kinematic equation $vdv = adx$.

4.1 Problem 4/25 (Conservation of Momentum)

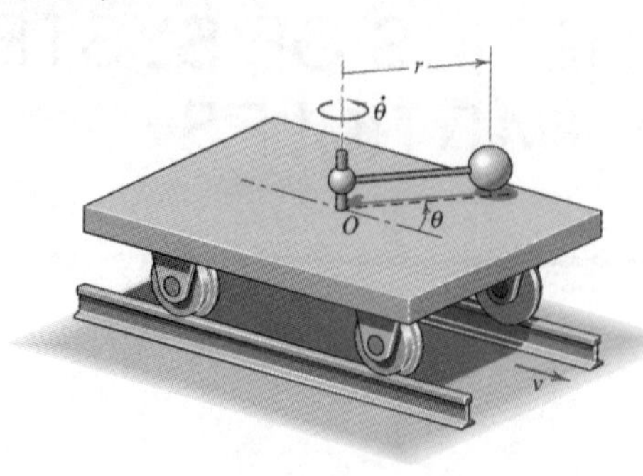

The small car, which has a mass of 20 kg, rolls freely on the horizontal track and carries the 5-kg sphere mounted on the light rotating rod with r = 0.4 m. A geared motor drive maintains a constant angular speed $\dot{\theta} = 4$ rad/s of the rod. If the car has a velocity v = 0.6 m/s when $\theta = 0$, plot v as a function of θ for one revolution of the rod. Also plot the absolute position of the sphere for two revolutions of the rod. Neglect the mass of the wheels and any friction.

Problem Formulation

Since $\Sigma F_x = 0$ we have conservation of momentum in the x direction. The diagram to the right shows the system at $\theta = 0$ and at an arbitrary angle θ. From the relative velocity equation, the velocity of the sphere is the vector sum of the velocity of the car (v) and the velocity of the sphere relative to the car ($r\dot{\theta}$).

$$(G_x)_{\theta=0} = 20(0.6) + 5(0.6) = 15 \text{ N}\bullet\text{s}$$

$$(G_x)_\theta = 20v + 5\left(v - r\dot{\theta}\sin\theta\right) = 25v - 8\sin\theta$$

Setting $(G_x)_{\theta=0} = (G_x)_\theta$ and solving yields,

$$v = 0.6 + 0.32\sin\theta$$

Now let time $t = 0$ be the time when $\theta = 0$ and place an x-y coordinate system at the center of the car as shown in the diagram so that $x(t)$ is the position of the center of the car. Since $v = dx/dt$ and $\theta = 4t$ we have,

$$x = \int_0^t v dt = \int_0^t \left(0.6 + 0.32\sin(4t)\right)dt = 0.6t + 0.08\left(1 - \cos(4t)\right)$$

The x and y components of the sphere can now be determined as,

$$x_s = x + r\cos\theta = 0.08 + 0.6t + 0.32\cos(4t)$$

$$y_s = r\sin\theta = 0.4\sin(4t)$$

The absolute position of the sphere can be obtained by plotting y_s versus x_s. The time required for two revolutions of the arm is $4\pi/4 = \pi$ seconds.

MATLAB Scripts

```
%%%%%%%%%%%%%%%% Script #1 %%%%%%%%%%%%%%%%%
% This script plots v as a function of theta
theta = 0:0.02:2*pi;
v = 0.6 + 0.32*sin(theta);
plot(theta*180/pi, v)
xlabel('theta (degrees)')
ylabel('v (m/s)')
%%%%%%%%%%%%%%% end of script %%%%%%%%%%%%%%
```

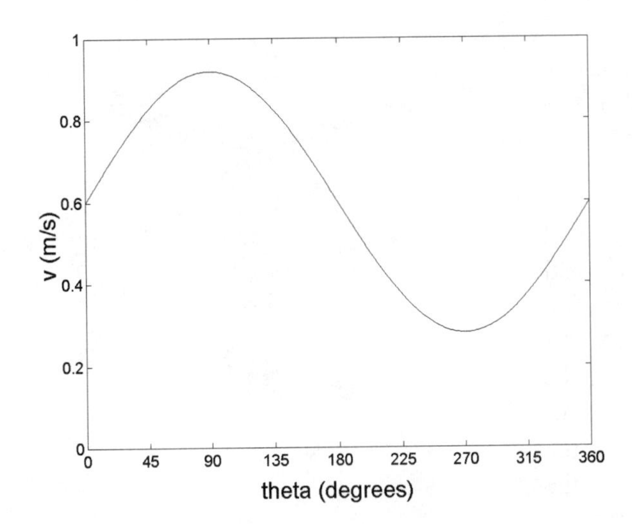


```
%%%%%%%%%%%%%%%% Script #2 %%%%%%%%%%%%%%%%%
% This script plots the position of the sphere
t = 0:0.01:pi;
xs = 0.08+0.6*t+0.32*cos(4*t);
ys = 0.4*sin(4*t);
plot(xs, ys)
xlabel('x (m)')
ylabel('y (m)')
title('position of the sphere')
%%%%%%%%%%%%%%% end of script %%%%%%%%%%%%%%
```

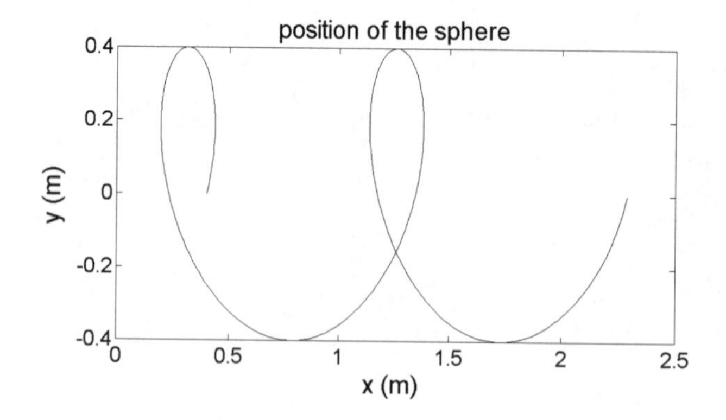

4.2 Problem 4/58 (Steady Mass Flow)

The sprinkler is made to rotate at the constant angular velocity ω and distributes water at the volume rate Q. Each of the four nozzles has an exit area A. Write an expression for the torque M on the shaft of the sprinkler necessary to maintain the given motion. Here we would like to study the effects of the geometry of the sprinkler upon this torque. To this end, it is helpful to introduce the non-dimensional parameters $M' = M/4\rho A r u^2$, $\Omega = \omega r/u$, and $\beta = b/r$ where $u = Q/4A$ is the velocity of the water relative to the nozzle and ρ is the density of the water. Plot the non-dimensional torque M' versus β for $\Omega = 0.5$, 1, and 2. Let Ω_0 be the non-dimensional velocity Ω at which the sprinkler will operate with no applied torque. Plot Ω_0 versus β. For both plots let β range between 0 and 1.

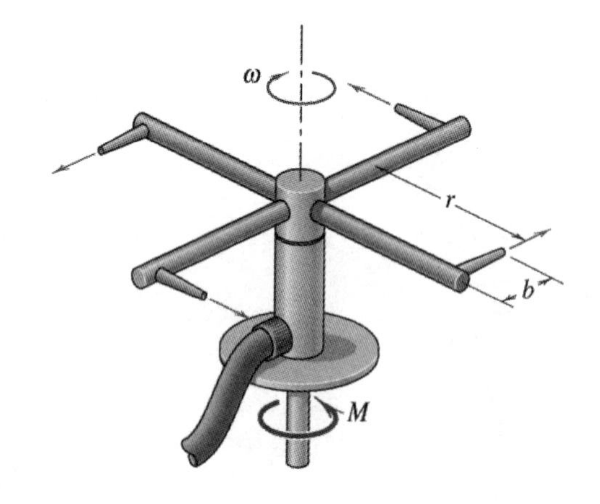

Problem Formulation

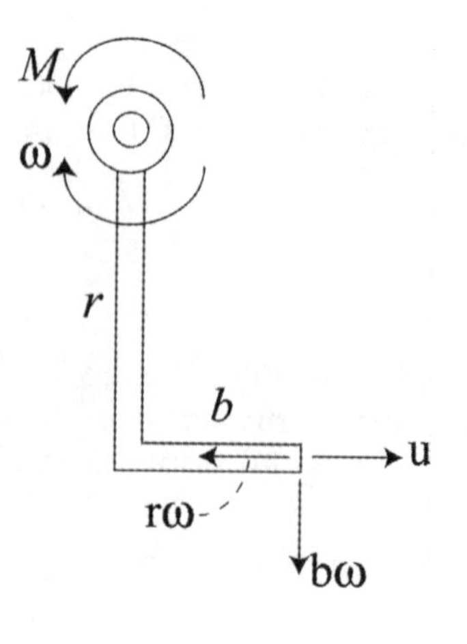

The figure to the right shows the three components of the absolute velocity of the water at the exit. u (= $Q/4A$) is the velocity of the water relative to the nozzle. The mass flow rate $m' = \rho Q$. Taking clockwise as positive, the application of equation 4/19 of your text yields,

$$\Sigma M_0 = -M = \rho Q\left(\omega r^2 + \omega b^2 - ur - 0\right)$$

$$M = \rho Q\left(ur - \omega\left(r^2 + b^2\right)\right)$$

Now we want to introduce the non-dimensional parameters defined in the problem statement. For many undergraduate students, non-dimensional analysis is a very confusing topic. It is important to realize that the difficulty is really that of determining which non-dimensional parameters are appropriate for a particular problem. If these parameters have already been defined, as in this problem, all you have to do is substitute. In this case we merely substitute $M = 4\rho Aru^2 M'$, $\omega = \Omega u/r$, and $b = r\beta$ into the equation above. When this is done many terms will cancel yielding,

$$M' = 1 - \Omega\left(1 + \beta^2\right)$$

Setting $M' = 0$ we can solve for Ω_0,

$$\Omega_0 = \frac{1}{1 + \beta^2}$$

MATLAB Scripts

```
%%%%%%%%%%%%%%%% Script #1 %%%%%%%%%%%%%%%%
% This script plots the non-dimensional torque
% M' as a function of beta = b/r for Omega =
% 0.5, 1, and 2
beta = 0:0.01:1;
Mp = inline('1-Omega*(1+beta.^2)')
plot(beta, Mp(0.5,beta),beta, Mp(1,beta),beta, Mp(2,beta))
xlabel('beta = b/r')
title('non-dimensional torque')
%%%%%%%%%%%%% end of script %%%%%%%%%%%%%%%%
```

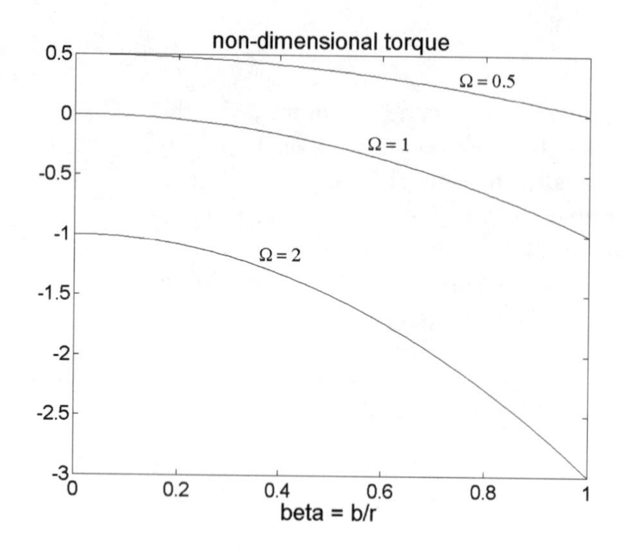


```
%%%%%%%%%%%%%%% Script #2 %%%%%%%%%%%%%%%
% This script plots the non-dimensional angular
% velocity Omega0 as a function of beta = b/r
beta = 0:0.01:1;
Omega0 = 1./(1+beta.^2);
plot(beta, Omega0)
xlabel('beta = b/r')
%%%%%%%%%%%% end of script %%%%%%%%%%%%%%%
```

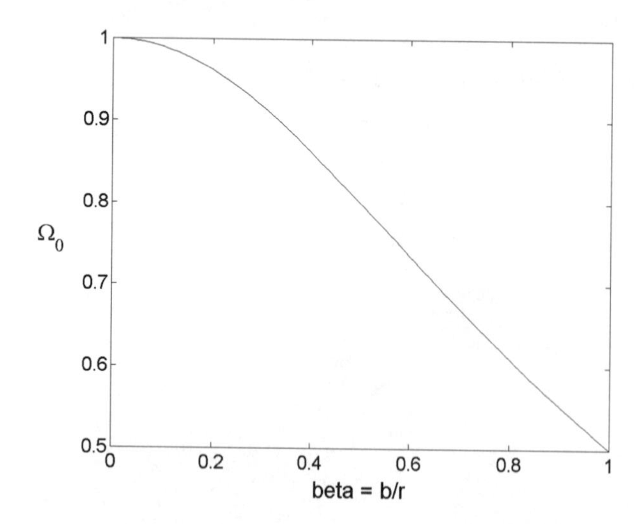

4.3 Problem 4/82 (Variable Mass)

The open-link chain of length L and mass ρ per unit length is released from rest in the position shown, where the bottom link is almost touching the platform and the horizontal section is supported on a smooth surface. Friction at the corner guide is negligible. Determine (a) the velocity v_1 of end A as it reaches the corner and (b) its velocity v_2 as it strikes the platform. Plot v_1 and v_2 as functions of h for $L = 5$ m.

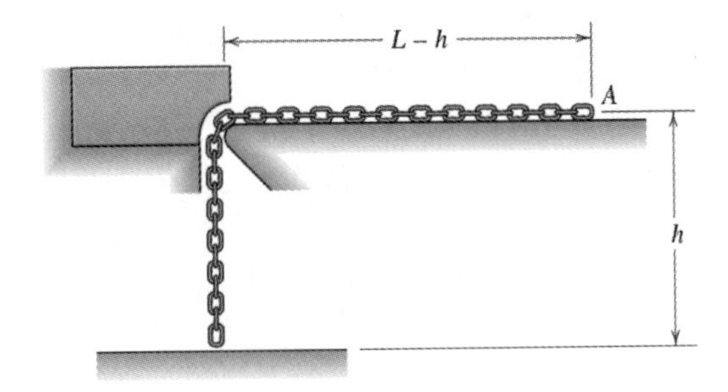

Problem Formulation

Let x be the displacement of the chain and T be the tension in the chain at the corner as shown in the diagram to the right. Note that the acceleration of the horizontal and vertical sections are both equal to $\ddot{x}$.

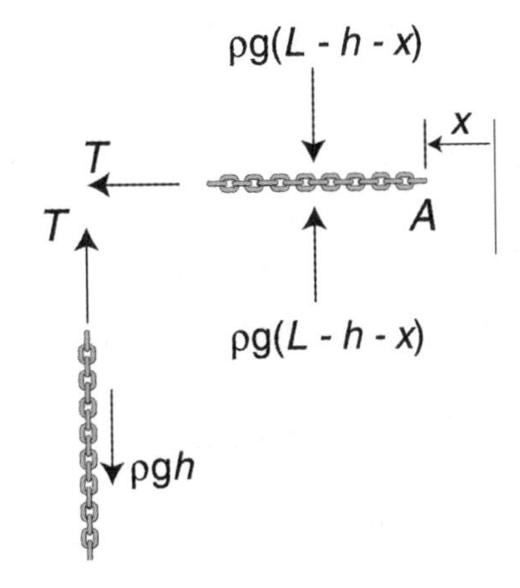

For the horizontal section,

$$[\Sigma F_x = ma_x] \qquad T = \rho(L-h-x)\ddot{x}$$

For the vertical section,

$$\downarrow [\Sigma F_y = ma_y] \qquad \rho gh - T = \rho h\ddot{x}$$

Substituting T from the first equation into the second and simplifying gives,

$$\ddot{x} = \frac{gh}{L-x}$$

Now we use the relation $vdv = adx$ to write,

$$\int_0^{v_1} vdv = \int_0^{L-h} \frac{gh}{L-x}dx \qquad v_1^2 = 2\int_0^{L-h} \frac{gh}{L-x}dx = 2gh\ln\left(\frac{L}{h}\right)$$

$$v_1 = \sqrt{2gh\ln(L/h)}$$

After end A has passed the corner it will be in free-fall until it hits the platform. With y positive down we have $vdv = gdy$ yielding,

$$\frac{1}{2}\left(v_2^2 - v_1^2\right) = gh$$

Substituting for v_1 and solving,

$$v_2 = \sqrt{2gh(1+\ln(L/h))}$$

MATLAB Worksheet and Script

```
EDU» syms v v1 v2 h g x L
```

First we solve the equation $\int_0^{v_1} vdv - \int_0^{L-h} gh/(L-x)dx = 0$ for v_1, omitting (as usual) the "=0".

```
EDU» eqn1 = int(v,v,0,v1)-int(g*h/(L-x),x,0,L-h)
eqn1 =
1/2*v1^2+log(h)*g*h-log(L)*g*h

EDU» solve(eqn1,v1)
ans =
[ (-2*log(h)*g*h+2*log(L)*g*h)^(1/2)]
[ -(-2*log(h)*g*h+2*log(L)*g*h)^(1/2)]
```

MATLAB has found two solutions. The first is the one we want since it is positive. This solution can easily be simplified to the result given above in the problem formulation section. Once v_1 is known it is rather easy to find v_2. We'll do it symbolically here for purposes of illustration.

First we copy and paste the first solution above to define v_1. Then we solve the equation $\int_{v_1}^{v_2} vdv - \int_0^h gdy = 0$ for v_2. Note how the result for v_1 is automatically substituted.

```
EDU» v1 = (-2*log(h)*g*h+2*log(L)*g*h)^(1/2);
EDU» eqn2 = int(v,v,v1,v2)-int(g,x,0,h)
eqn2 =
1/2*v2^2+log(h)*g*h-log(L)*g*h-g*h
```

```
EDU» solve(eqn2,v2)
ans =
[ (-2*log(h)*g*h+2*log(L)*g*h+2*g*h)^(1/2)]
[ -(-2*log(h)*g*h+2*log(L)*g*h+2*g*h)^(1/2)]
```

Once again, the first solution will simplify to that given in the problem formulation section above.

```
%%%%%%%%%%%%%%%%%%% Script %%%%%%%%%%%%%%%%%%%%
% This script plots v1 and v2 as functions of
% h for L = 5 m
L = 5; g = 9.81;
h = 0:0.01:L;
v1 = sqrt(2*g*h.*log(L./h));
v2 = sqrt(2*g*h.*(1+log(L./h)));
plot(h, v1, h, v2)
xlabel('h (m)')
ylabel('velocity (m/s)')
%%%%%%%%%%%%%%% end of script %%%%%%%%%%%%%%%%%
```

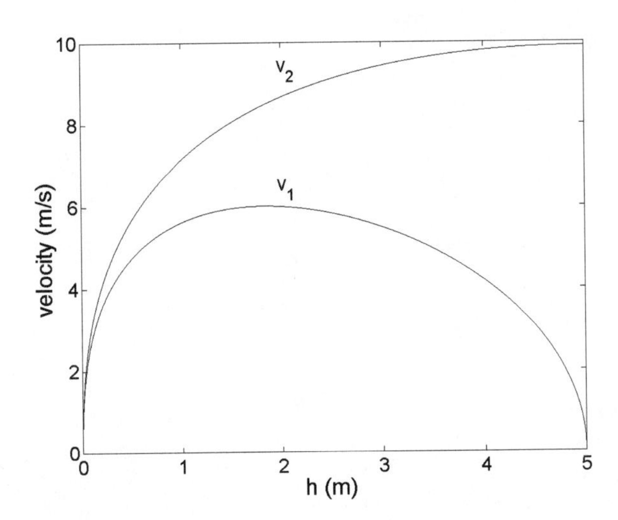

PLANE KINEMATICS OF RIGID BODIES

5

This chapter extends the kinematic analysis of particles covered in chapter 2 to rigid bodies by taking into account the rotational motion of the body. Thus, the motion of rigid bodies involves both translation and rotation. Problem 5.1 is a straightforward problem where the angular position (θ) of a line on a rotating disk is given as a function of time. *diff* is used to differentiate θ twice yielding first the angular velocity and then the angular acceleration. *solve* is then used to find the time at which the angular acceleration is zero. Problem 5.2 is an interesting application of absolute motion analysis. The problem illustrates the usefulness of non-dimensional analysis in conducting a parametric study of the effects of geometry upon the performance of a hydraulic lift. In problem 5.3 the velocity of the piston in a reciprocating engine is plotted versus the angular orientation of the crank. The maximum velocity and the corresponding orientation (of the crank) are obtained using *diff* and *solve*. The instantaneous center of zero velocity is used in problem 5.4 to relate the velocity of the center of a disk to its angular velocity and to the angular velocity of a lever in no-slip contact with the disk. The problem illustrates very well the timesavings resulting from the use of computer algebra software. Problem 5.5 considers the reciprocating engine of problem 5.3 but uses absolute rather than relative motion analysis. The problem illustrates the use of computer algebra for carrying out some tedious calculations including differentiation and substitution.

5.1 *Problem 5/12 (Rotation)*

The angular position of a radial line on a rotating disk is given by $\theta = \left(-1+\frac{3}{2}t\right)e^{-t/2}$, where θ is in radians and t is in seconds. Plot the angular position, angular velocity, and angular acceleration versus time for the first 20 seconds of motion. Determine the time at which the acceleration is zero.

Problem Formulation

With θ defined explicitly as a function of time it is easy to find ω and α from their definitions,

$$\omega = \frac{d\theta}{dt} \qquad \alpha = \frac{d\omega}{dt}$$

The derivatives will be evaluated in the worksheet below. Once α is known as a function of time we solve the equation $\alpha = 0$ for t.

MATLAB Worksheet and Scripts

```
EDU» syms t
EDU» theta = (-1+3/2*t)*exp(-t/2)
theta =
(-1+3/2*t)*exp(-1/2*t)

EDU» omega = diff(theta,t)
omega =
3/2*exp(-1/2*t)-1/2*(-1+3/2*t)*exp(-1/2*t)

EDU» alpha = diff(omega,t)
alpha =
-3/2*exp(-1/2*t)+1/4*(-1+3/2*t)*exp(-1/2*t)

EDU» solve(alpha,t)
ans = 14/3
```

Remember that the command solve(alpha, t) solves the equation alpha = 0 for t. The "= 0" is automatic. Thus, $\alpha = 0$ when $t = 14/3 = 4.67$ sec.

The above results are copied and pasted into the following scripts for plotting. When doing this, don't forget to add "." at appropriate places to insure term by term rather than matrix operations.

```
%%%%%%%%%%%%%%%% Script #1 %%%%%%%%%%%%%%%%%%%%
% This script plots theta as a function of time
t = 0:0.1:20;
theta = (-1+3/2*t).*exp(-1/2*t);
plot(t, theta*180/pi)
xlabel('time (sec)')
title('angular position (degrees)')
%%%%%%%%%%%% end of script %%%%%%%%%%%%%%%%%%%%%
```

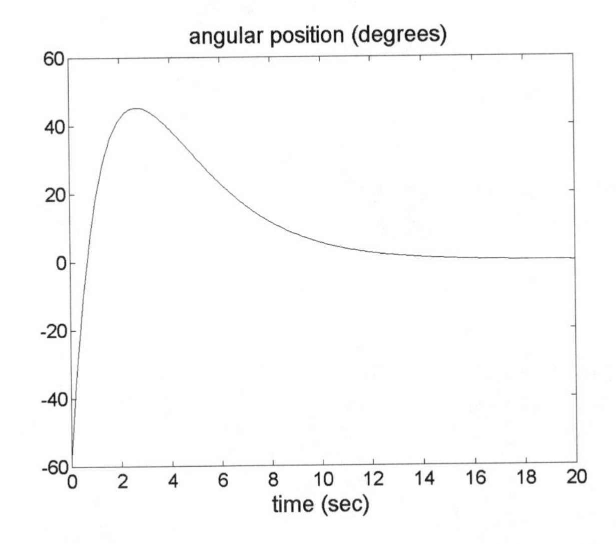


```
%%%%%%%%%%%%%%%% Script #2 %%%%%%%%%%%%%%%%%%%%
% This script plots omega as a function of time
t = 0:0.1:20;
omega = 3/2*exp(-1/2*t)-1/2*(-1+3/2*t).*exp(-1/2*t);
plot(t, omega)
xlabel('time (sec)')
title('angular velocity (rad/sec)')
%%%%%%%%%%%% end of script %%%%%%%%%%%%%%%%%%%%%
```

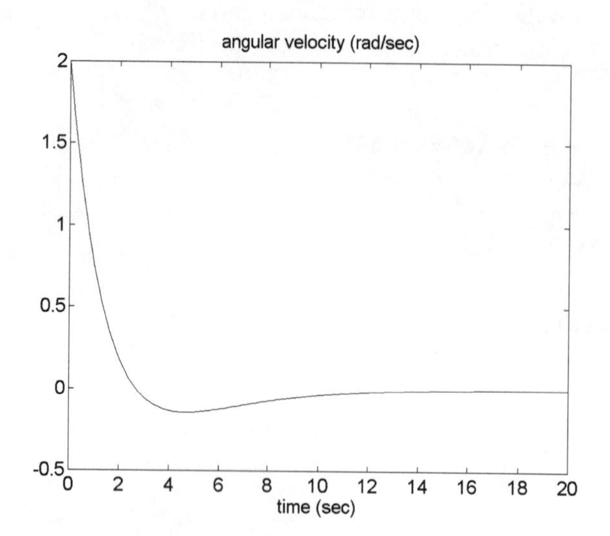


```
%%%%%%%%%%%%%%% Script #3 %%%%%%%%%%%%%%%%%%
% This script plots alpha as a function of time
t = 0:0.1:20;
alpha = -3/2*exp(-1/2*t)+1/4*(-1+3/2*t).*exp(-1/2*t);
plot(t, alpha)
xlabel('time (sec)')
title('angular acceleration (rad/s^2)')
%%%%%%%%%%%% end of script %%%%%%%%%%%%%%%%%%%
```

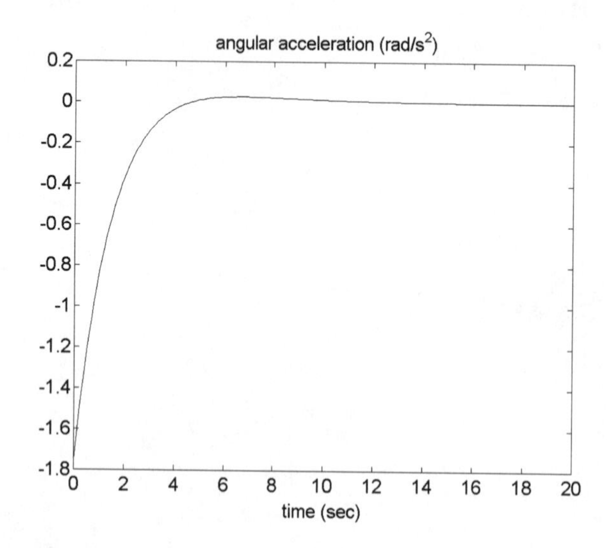

5.2 *Problem 5/39 (Absolute Motion)*

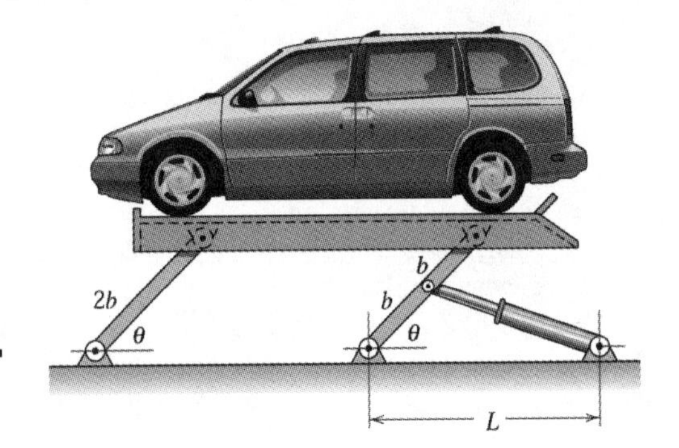

Derive an expression for the upward velocity v of the car hoist system in terms of θ. The piston rod of the hydraulic cylinder is extending at the rate $\dot{s}$. Plot the non-dimensional velocity $v/\dot{s}$ as a function of θ for $b/L = 0.1, 0.5, 1$, and 2.

Problem Formulation

From the diagram to the right,

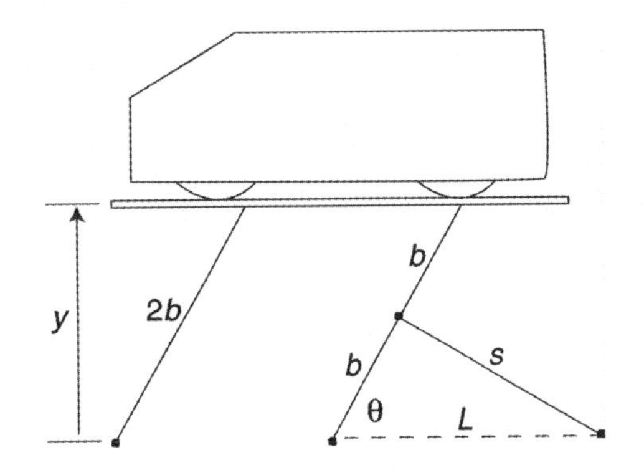

$$y = 2b\sin\theta$$

$$\dot{y} = 2b\dot{\theta}\cos\theta$$

If the angular velocity $\dot{\theta}$ were known as a function of θ we would be finished. The motion of the car hoist system is controlled by the extension rate $\dot{s}$ of the hydraulic cylinder rather than the angular velocity. Thus, to complete the problem we need to relate $\dot{\theta}$ and $\dot{s}$.

$$s^2 = L^2 + b^2 - 2Lb\cos\theta$$

$$2s\dot{s} = 0 + 0 + 2Lb\dot{\theta}\sin\theta$$

$$\dot{\theta} = \frac{s\dot{s}}{Lb\sin\theta}$$

Substituting,

$$v = \frac{2bs\dot{s}}{Lb\sin\theta}\cos\theta = \frac{2\dot{s}\sqrt{L^2 + b^2 - 2Lb\cos\theta}}{L\tan\theta}$$

$$v/\dot{s} = \frac{2\sqrt{1 + (b/L)^2 - 2(b/L)\cos\theta}}{\tan\theta} = \frac{2\sqrt{1 + \beta^2 - 2\beta\cos\theta}}{\tan\theta}$$

where $\beta = b/L$.

MATLAB Script

```
%%%%%%%%%%%%%%%%%%%%%%%%%%%%%%%%%%%%%%%%%%%%%%%%%%%%%%%%%
% v is the non-dimensional velocity
v = inline('2/tan(theta)*sqrt(1^2+beta^2-2*beta*cos(theta))')
v = vectorize(v);
% the above adds "." to insure term by term rather than
% matrix operations
th = 10*pi/180:0.01:pi/2;
deg = th*180/pi;
plot(deg,v(.1,th),deg,v(.5,th),deg,v(1,th),deg,v(2,th))
axis([10 90 0 5])
xlabel('theta (degrees)')
title('non-dimensional velocity')
%%%%%%%%%%%%%%%%%%%%%%%%%%%%%%%%%%%%%%%%%%%%%%%%%%%%%%%%%
```

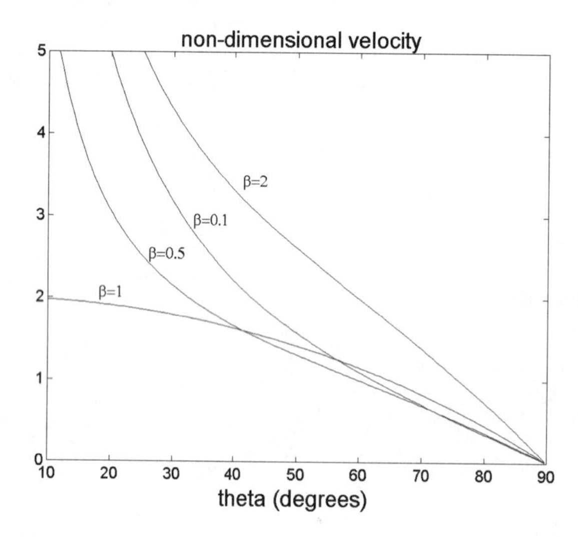

5.3 Sample Problem 5/9 (Relative Velocity)

The common configuration of a reciprocating engine is that of the slider crank mechanism shown. If crank OB has a clockwise rotational speed of 1500 rev/min; (a) Plot v_A versus θ for one revolution of the crank. (b) Find the maximum speed of the piston A and the corresponding value of θ.

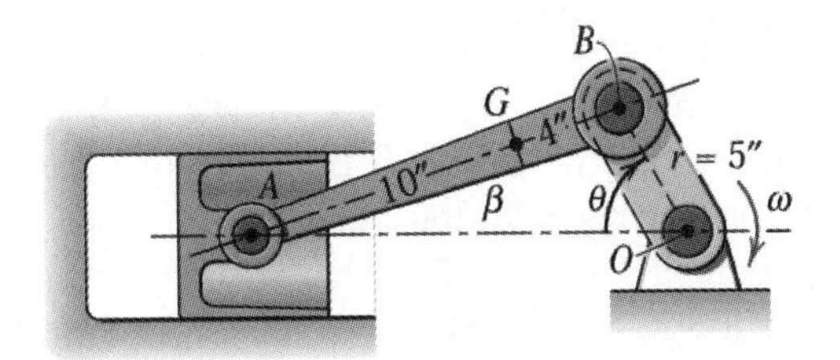

Problem Formulation

Let l be the length of connecting rod AB. Start with the relative velocity equation

$$\mathbf{v}_B = \mathbf{v}_A + \mathbf{v}_{B/A}$$

The crank pin velocity is $v_B = r\omega$ and is normal to OB. The velocity of A is horizontal while the velocity of B/A has magnitude $l\omega_{AB}$ and is directed perpendicular to AB. The angle β can be found in terms of θ by using the law of sines,

$$\sin\beta = \frac{r}{l}\sin\theta \quad \text{Also, } \cos\beta = \sqrt{1-\sin^2\beta} = \sqrt{1-\frac{r^2}{l^2}\sin^2\theta}$$

From the vector diagram to the right,

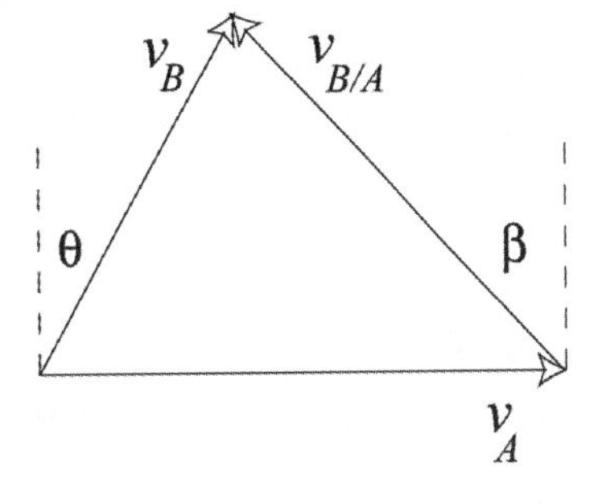

$$v_B\cos\theta = v_{B/A}\cos\beta\text{. Thus, } v_{B/A} = \frac{\cos\theta}{\cos\beta}r\omega$$

Also from the diagram, $v_A = v_B\sin\theta + v_{B/A}\sin\beta$. Substitution yields

$$v_A = r\omega(\sin\theta + \cos\theta\tan\beta) = r\omega\sin\theta\left(1+\frac{r\cos\theta}{l\sqrt{1-\frac{r^2}{l^2}\sin^2\theta}}\right)$$

Note that v_A has been expressed explicitly in terms of θ by substituting for $\cos\beta$ and $\tan\beta$. This has been done only for sake of clarity. When working with a computer such substitutions will be automatic once β has been defined in terms of θ ($\beta = \sin^{-1}(r\sin\theta / l)$).

(b) The angle θ at which the maximum value of v_A occurs is found by solving the equation $dv_A/d\theta = 0$ for θ. Evaluating this derivative and solving the resulting equation for θ would be difficult without the help of a computer. It turns out that there are many solutions to this equation, most of which are complex. In the results that follow we find that the maximum occurs at $\theta = 1.261$ radians (72.3°). Substitution of this result into the expression for v_A yields the maximum speed of the piston A, $v_A = 69.6$ ft/sec.

MATLAB Worksheet and Scripts

```
%%%%%%%%%%%%%%% Script #1 %%%%%%%%%%%%
% This script solves for vA using symbolic
% algebra
syms r L theta omega
vB = r*omega;
beta = asin(r/L*sin(theta));
vBA = cos(theta)/cos(beta)*r*omega;
vA = vB*sin(theta)+vBA*sin(beta)
%%%%%%%%%%%%% end of script %%%%%%%%%%%
```

Output of script #1

vA =
r*omega*sin(theta)+cos(theta)/(1-r^2/L^2*sin(theta)^2)^(1/2)*
r^2*omega/L*sin(theta)

```
%%%%%%%%%%%%%%%%%%% Script #2 %%%%%%%%%%%%%%%%%%%
% This script plots the velocity of piston A
% as a function of theta
r = 5/12;
L = 14/12;
o = 1500*2*pi/60; % o=omega
x = 0:0.01:2*pi;   % x=theta
vA = r*o*sin(x)+cos(x)./(1-r^2/L^2*sin(x).^2).^(1/2)*r^2*o/L.*sin(x);
plot(x*180/pi,vA)
xlabel('theta (degrees)')
title('velocity of piston A (ft/sec)')
%%%%%%%%%%%%% end of script %%%%%%%%%%%
```

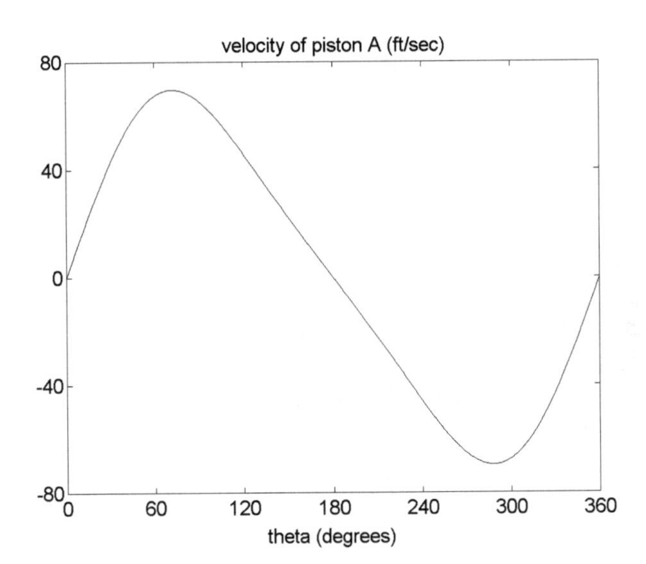

Part (b)

```
%%%%%%%%%%%%%%%%%%% Script #3 %%%%%%%%%%%%%
% This script solves for the values of theta which
% make vA a maximum
r = 5/12;
L = 14/12;
omega = 1500*2*pi/60;
syms theta
% The following is copied and pasted from the output of
% script #1
vA = r*omega*sin(theta)+cos(theta)/(1-
r^2/L^2*sin(theta)^2)^(1/2)*r^2*omega/L*sin(theta)
dvAdtheta = diff(vA,theta)
solve(dvAdtheta,theta)
%%%%%%%%%%%%%%%%%%% end of script %%%%%%%%%%%
```

Output of script #3

vA =
125/6*pi*sin(theta)+625/84*cos(theta)/(1-25/196*sin(theta)^2)^(1/2)*
pi*sin(theta)

dvAdtheta =
125/6*pi*cos(theta)-625/84*sin(theta)^2/(1-25/196*sin(theta)^2)^(1/2)*pi+
15625/16464*cos(theta)^2/(1-25/196*sin(theta)^2)^(3/2)*pi*sin(theta)^2+
625/84*cos(theta)^2/(1-25/196*sin(theta)^2)^(1/2)*pi

```
ans =

[ 3.14159 + 1.5852*i]
[ 3.14159 -  1.5852*i]
[                 1.261204]
[  -1.5708 + 1.9311*i]
[    1.5708 - 1.9311*i]
[                -1.261204]
```

We see from the above that MATLAB has found six solutions. The only two of these that are not complex are $\theta = \pm$ 1.261 radians ($\pm$ 72.3°). From the plot, the first of these is clearly the maximum. After running the script above, you can find the maximum speed by substitution as follows.

```
EDU» subs(vA,theta,1.2612)

ans = 69.5514
```

Thus, the maximum speed of the piston is v_A = 69.6 ft/sec at θ = 72.3°.

5.4 Problem 5/110 (Instantaneous Center)

Rotation of the lever OA is controlled by the motion of the contracting circular disk whose center is given a horizontal velocity v. Derive an expression for the angular velocity ω_{OA} of lever OA and the angular velocity ω_w of the wheel in terms of x. At what value of x will the two angular velocities have the same magnitude? Plot the angular velocities as functions of x if $r = 1$ m and $v = 2$ m/s. Let x range between 1 and 2 meters and restrict the vertical axis to be between 0 and 3. You may assume that the wheel does not slip on the lever.

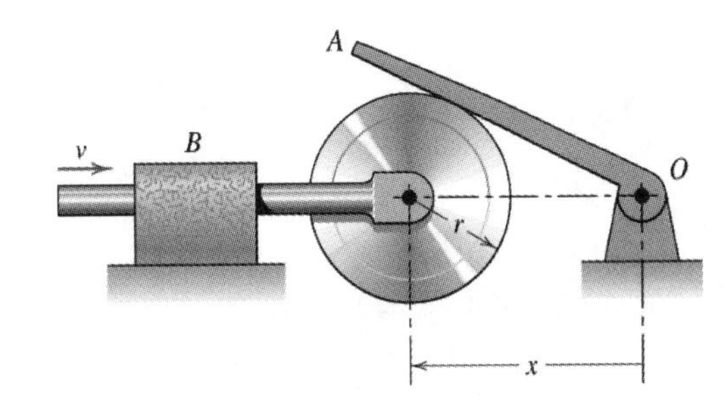

Problem Formulation

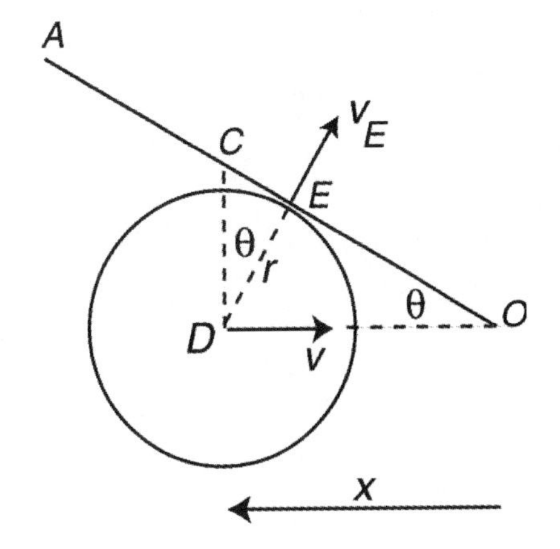

The instantaneous center of zero velocity for the wheel is shown (point C) in the diagram to the right. First we define the lengths required for the velocity analysis.

$$CD = r / \cos\theta \qquad CE = r\tan\theta \qquad EO = x\cos\theta$$

Using the instantaneous center we can now write,

$$\omega_w = \frac{v}{CD}$$

$$\omega_{OA} = \frac{v_E}{EO} = \frac{1}{EO} v \frac{CE}{CD} = \frac{CE}{EO\ CD} v$$

Substituting CD, CE, and EO gives the angular velocities as functions of θ. Unfortunately, the problem asks for these velocities as functions of x. If we were doing this problem "by hand" we would now have to write the trig functions $\sin\theta$, $\cos\theta$, and $\tan\theta$ in terms of x and then substitute and simplify. This could obviously become rather tedious. One advantage to using a computer algebra program is that substitutions such as these will be automatic. All you have to do is start by defining θ as a function of x,

$$\theta = \sin^{-1}(r / x)$$

Now, as you type in the analysis above, the program will automatically substitute for θ.

MATLAB Worksheet and Script

```
EDU» syms v x r theta
EDU» theta = asin(r/x)
theta  = asin(r/x)

EDU» CD = r/cos(theta)
CD  = r/(1-r^2/x^2)^(1/2)

EDU» CE = r*tan(theta)
CE  = r^2/x/(1-r^2/x^2)^(1/2)

EDU» EO = x*cos(theta)
EO  = x*(1-r^2/x^2)^(1/2)

EDU» omega_w = v/CD
omega_w  =  v/r*(1-r^2/x^2)^(1/2)

EDU» omega_OA = v*CE/EO/CD
omega_OA  = v*r/x^2/(1-r^2/x^2)^(1/2)
```

The expressions for the two angular velocities are copied and pasted into the following script so that they can be plotted as functions of x. When doing this, be sure to remember to put periods "." at appropriate places to insure term by term rather than matrix operations.

Now we want to find the value of x where the two angular velocities are equal. For this we will use the *solve* command. When using the solve command we need to re-write the equation in the form *expression* = 0 and then omit the "=0". In other words we solve the equation $\omega_w - \omega_{OA} = 0$ instead of $\omega_w = \omega_{OA}$.

```
EDU» solve(omega_w - omega_OA,x)
ans =
[ 2^(1/2)*r]
[ -2^(1/2)*r]
```

Thus, $\omega_w = \omega_{OA}$ when $x = \sqrt{2}r$.

```
%%%%%%%%%%%%%%%%%%%% Script %%%%%%%%%%%%%%%%
% This script plots the two angular velocities
% as functions of x
v = 2; r = 1;
```

```
x = 1:0.01:2;
omega_OA = v*r./x.^2./(1-r^2./x.^2).^(1/2);
omega_w = v/r*(1-r^2./x.^2).^(1./2);
plot(x, omega_OA, x, omega_w);
axis([1 2 0 3])
xlabel('x (m)')
title('angular velocity (rad/s)')
%%%%%%%%%%%%%%% end of script %%%%%%%%%%%%%%%
```

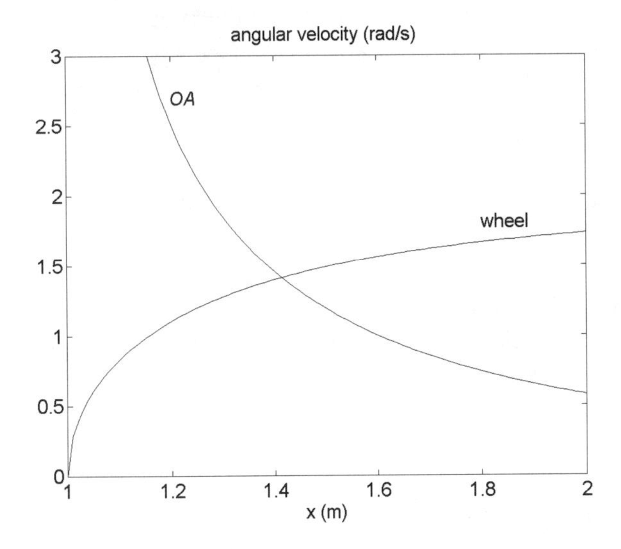

5.5 *Sample Problem 5/15 (Absolute Motion)*

The common configuration of a reciprocating engine is that of the slider crank mechanism shown. If crank OB has a clockwise rotational speed of 1500 rev/min; (a) Plot v_A and v_G versus time for two revolutions of the crank. (b) Plot a_A and a_G versus time for two revolutions of the crank.

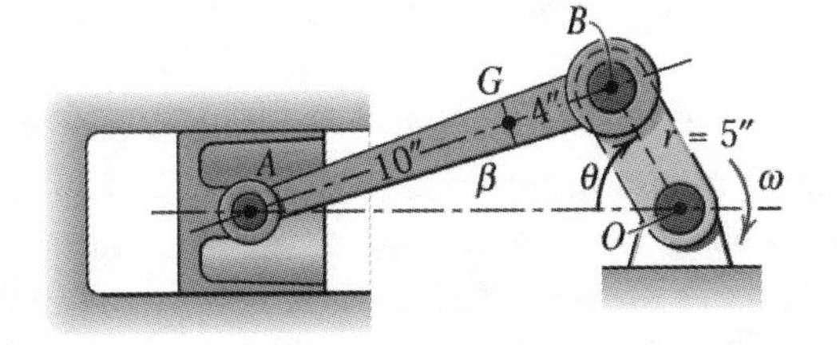

Problem Formulation

This problem appears in sample problems 5/9 and 5/15 in your text. Sample problem 5/9 considers a relative velocity analysis while sample problem 5/15 uses a relative acceleration analysis. Generally speaking, the easiest approach to use with a computer is an absolute motion analysis, provided you have software capable of doing symbolic algebra and calculus such as MATLAB. We will use the present problem to illustrate this approach.

We start by using the law of sines ($\sin(\beta)/r = \sin(\theta)/l$) to express β as a function of θ

$$\beta = \sin^{-1}\left(\frac{r}{l}\sin\theta\right)$$

where l is the length of connecting rod AB and β is the angle between AB and the horizontal. Now place an x-y coordinate system at O with x positive to the right and y positive up and write expressions for the coordinates of A and G in terms of θ and β

$$x_A = -r\cos\theta - l\cos\beta$$

$$x_G = -r\cos\theta - \bar{r}\cos\beta \qquad y_G = (l - \bar{r})\sin\beta$$

where $\bar{r}$ is the distance from B to G (4 in. in the figure). All that is needed to find the velocities v_A, v_{Gx}, and v_{Gy} is to differentiate these expressions with respect to time. The magnitude of the velocity of G is then found from

$$v_G = \sqrt{{v_{Gx}}^2 + {v_{Gy}}^2}$$

The accelerations a_A, a_{Gx}, and a_{Gy} are then found by differentiating v_A, v_{Gx}, and v_{Gy} with the magnitude of the acceleration of G being obtained from,

$$a_G = \sqrt{{a_{Gx}}^2 + {a_{Gy}}^2}$$

Since we will be differentiating with respect to time, the first thing we will do in the computer program is to define θ as a function of time. Then, when we write the above expressions for β, x_A, x_G, and y_G, the computer will automatically substitute for θ rendering each of these as functions of time. Assuming that θ is initially zero,

$$\theta(t) = \omega t = \frac{1500(2\pi)}{60}t = 157.1t$$

The problem statement asks us to plot versus time for two revolutions ($\theta = 4\pi$ radians) of the crank. The time required for two revolutions is $4\pi/157.1 = 0.08$ sec.

MATLAB Worksheet and Scripts

```
%%%%%%%%%%%%% Script #1 %%%%%%%%%%%%%%%%%%
% This script uses symbolic algebra to find
% the velocity and acceleration of A and G
% w = constant angular velocity
% rb = r_bar (4 in. in the figure)
syms r rb L w t
theta = w*t;
beta = asin(r/L*sin(theta));
xA = -r*cos(theta)-L*cos(beta)
xG = -r*cos(theta)-rb*cos(beta)
yG = (L-rb)*sin(beta)
vA = diff(xA,t)
vGx = diff(xG,t)
vGy = diff(yG,t)
vG = sqrt(vGx^2+vGy^2)
aA = diff(vA,t)
aGx = diff(vGx,t)
aGy = diff(vGy,t)
aG = sqrt(aGx^2+aGy^2)
%%%%%%%%%%%%%%%%% end of script %%%%%%%%%%%%%%%%%
```

Output of Script #1

xA = -r*cos(w*t)-L*(1-r^2/L^2*sin(w*t)^2)^(1/2)

xG = -r*cos(w*t)-rb*(1-r^2/L^2*sin(w*t)^2)^(1/2)

yG = (L-rb)*r/L*sin(w*t)

vA =
r*sin(w*t)*w+1/L/(1-r^2/L^2*sin(w*t)^2)^(1/2)*r^2*sin(w*t)*cos(w*t)*w

vG =
((r*sin(w*t)*w+rb/(1-r^2/L^2*sin(w*t)^2)^(1/2)*r^2/L^2*
sin(w*t)*cos(w*t)*w)^2+(L-rb)^2*r^2/L^2*cos(w*t)^2*w^2)^(1/2)

aA =
r*cos(w*t)*w^2+1/L^3/(1-r^2/L^2*sin(w*t)^2)^(3/2)*r^4*sin(w*t)^2*
cos(w*t)^2*w^2+1/L/(1-r^2/L^2*sin(w*t)^2)^(1/2)*r^2*cos(w*t)^2*w^2-
1/L/(1-r^2/L^2*sin(w*t)^2)^(1/2)*r^2*sin(w*t)^2*w^2

aG =
((r*cos(w*t)*w^2+rb/(1-r^2/L^2*sin(w*t)^2)^(3/2)*r^4/L^4*sin(w*t)^2*
cos(w*t)^2*w^2+rb/(1-r^2/L^2*sin(w*t)^2)^(1/2)*r^2/L^2*cos(w*t)^2*w^2-

rb/(1-r^2/L^2*sin(w*t)^2)^(1/2)*r^2/L^2*sin(w*t)^2*w^2)^2+(L-rb)^2*
r^2/L^2*sin(w*t)^2*w^4)^(1/2)

The results for the velocities and accelerations will be used to produce the required plots in the following two scripts. The easiest thing to do is to copy the results from the worksheet into the scripts. Before doing this you should use the *vectorize* command to place periods at appropriate places in order to insure term by term rather than matrix operations. After running the script type "vectorize(vA)" in the worksheet and then copy and paste the result into the script. Repeat this process for vG, aA, and aG.

```
%%%%%%%%%%%%%%% Script #2 %%%%%%%%%%%%%%%%%%
% This script plots vA and vG versus time
r = 5/12;
rb = 4/12; % r_bar
L = 14/12;
w = 1500*2*pi/60; % w=omega
t = 0:0.001:0.08;
vA = r.*sin(w.*t).*w+1./L./(1-r.^2./L.^2.*sin(w.*t).^2).^
(1./2).*r.^2.*sin(w.*t).*cos(w.*t).*w;

vG = ((r.*sin(w.*t).*w+rb./(1-r.^2./L.^2.*sin(w.*t).^2).^
(1./2).*r.^2./L.^2.*sin(w.*t).*cos(w.*t).*w).^2+(L-
rb).^2.*r.^2./L.^2.*cos(w.*t).^2.*w.^2).^(1./2);
plot(t, vA, t, vG)
%%%%%%%%%%%%%%% end of script %%%%%%%%%%%%%%%
```

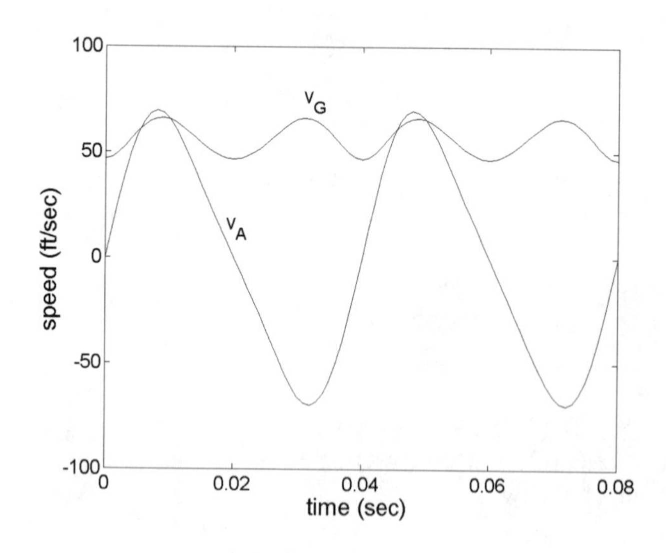

```
%%%%%%%%%%%%%%% Script #3 %%%%%%%%%%%%%%%%
% This script plots aA and aG versus time
r = 5/12;
rb = 4/12;
L = 14/12;
w = 1500*2*pi/60; % w=omega
t = 0:0.001:0.08;
aA = r.*cos(w.*t).*w.^2+1./L.^3./(1-r.^2./L.^2.*
sin(w.*t).^2).^(3./2).*r.^4.*sin(w.*t).^2.*cos(w.*t).^2.*w.
^2+1./L./(1-r.^2./L.^2.*sin(w.*t).^2).^(1./2).*r.^2.*
cos(w.*t).^2.*w.^2-1./L./(1-r.^2./L.^2.*sin(w.*t).^2).^
(1./2).*r.^2.*sin(w.*t).^2.*w.^2;

aG = ((r.*cos(w.*t).*w.^2+rb./(1-r.^2./L.^2.*
sin(w.*t).^2).^(3./2).*r.^4./L.^4.*sin(w.*t).^2.*cos(w.*t).
^2.*w.^2+rb./(1-r.^2./L.^2.*sin(w.*t).^2).^(1./2).
*r.^2./L.^2.*cos(w.*t).^2.*w.^2-rb./(1-r.^2./L.^2.*
sin(w.*t).^2).^(1./2).*r.^2./L.^2.*sin(w.*t).^2.*w.^2).^2+(
L-rb).^2.*r.^2./L.^2.*sin(w.*t).^2.*w.^4).^(1./2);
plot(t, aA, t, aG)
%%%%%%%%%%%%%% end of script %%%%%%%%%%%%%%
```

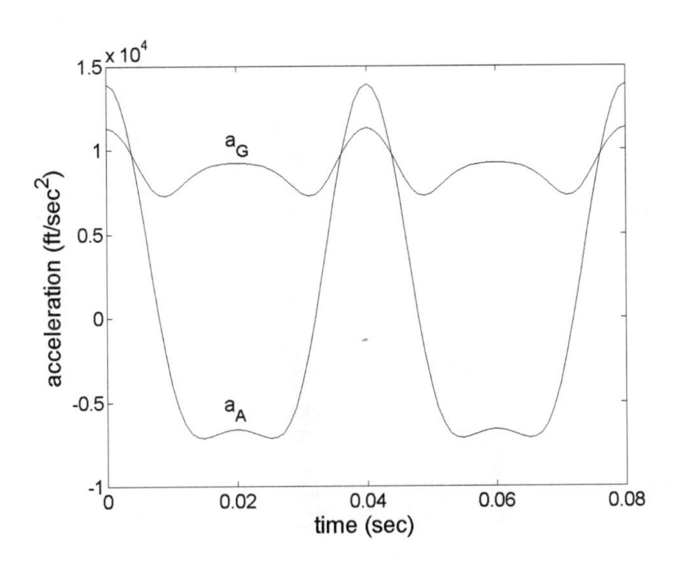

For further study

Suppose you wanted to solve this problem for the case where crank *OB* has a constant angular acceleration $\alpha = 60\ \text{rad/s}^2$. It turns out that you can solve this problem using exactly the same approach as above, changing only one line defining the dependence of θ upon time. Assuming that the system starts from rest at $\theta = 0$, the appropriate expression for θ is $\theta\ (t) = \tfrac{1}{2}\alpha t^2 = 30t^2$. The accelerations for this case are shown below in case you want to give it a try.

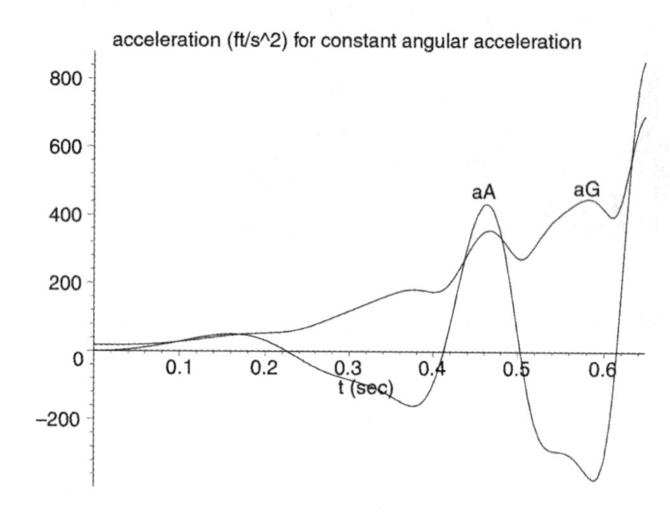

PLANE KINETICS OF RIGID BODIES

6

This chapter concerns the motion (translation and rotation) of rigid bodies that results from the action of unbalanced external forces and moments. In problem 6.1, three equations are solved for three unknown forces. The problem illustrates an alternative to the blunt (but straightforward) simultaneous solution of multiple equations. Instead, one of the equations is solved for one unknown after which the other two unknowns can be expressed directly in terms of that unknown. In this way, the results are immediately obtained via automatic substitution in MATLAB, thus avoiding some tedious algebra. The force at the hinge of a pendulum is plotted versus the angular position of the pendulum in problem 6.2. The algebra is rather simple in this case and MATLAB is used primarily for purposes of plotting. Problems 6.3 and 6.4 consider rigid bodies in general plane motion. *solve* is used in problem 6.3 to solve two equations simultaneously for two unknowns. The maximum acceleration of a point on the rigid body is then obtained by using *diff* and *solve*. In this problem, *solve* finds five solutions to the equation and it is necessary to determine which of these is physically correct. This problem is also interesting since a very natural "guess" of the value for the maximum acceleration turns out to be incorrect. Problem 6.4 is an example of a kinetics problem that also requires some kinematics. The angular acceleration of a bar is determined by summing moments. The kinematic equation $\omega d\omega = \alpha d\theta$ is then integrated to obtain the angular velocity. Problem 6.5 is an interesting work and energy problem that is complicated considerably by the fact that a spring is engaged for only part of the motion of a rotating bar. Symbolic algebra simplifies this problem considerably, though it is still rather tedious. It is common in Dynamics to find problems that require a combination of methods for their solution. Problem 6.6 is a good example involving both conservation of momentum and work/energy.

6.1 Problem 6/29 (Translation)

The parallelogram linkage is used to transfer crates from platform A to platform B and is hydraulically operated. The oil pressure in the cylinder is programmed to provide a smooth transition of motion from $\theta = 0$ to $\theta = \theta_0 = \pi/3$ rad given by $\theta = \frac{\pi}{6}\left(1 - \cos\frac{\pi t}{2}\right)$ where t is in seconds. Plot the magnitude of the forces at pins D and F versus θ for $0 \le \theta \le \pi/3$ (60°). The crate and the platform have a combined mass of 200 kg with mass center at G. The mass of each link is small and may be neglected.

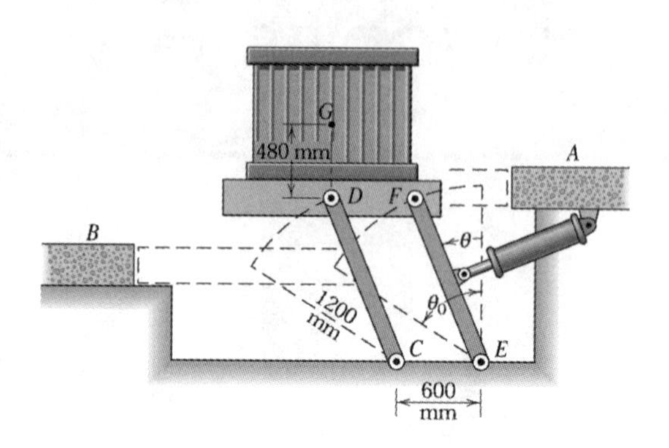

Problem Formulation

The free-body diagram (FBD) and mass acceleration diagram (MAD) are shown to the right. Since the crate and platform undergo curvilinear translation, the acceleration of all points on the crate and platform will be identical. Thus, we can obtain the acceleration of G immediately from that of point F which moves in a circular path about E,

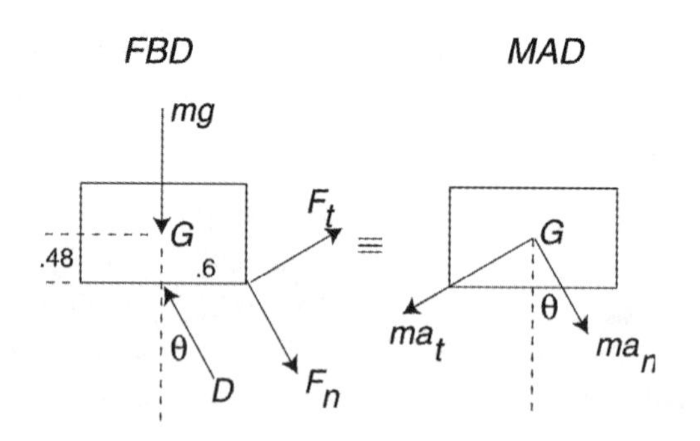

$$(\bar{a}_G)_n = a_n = r\dot{\theta}^2 = r\omega^2 \qquad (\bar{a}_G)_t = a_t = r\ddot{\theta} = r\alpha$$

where r = 1.2 m. To avoid possible confusion in what follows, be sure to remember that ω and α are the angular velocity and acceleration of the link. The crate and platform are not rotating. α and ω can be found from the definition of θ.

$$\theta = \frac{\pi}{6}\left(1 - \cos\frac{\pi t}{2}\right) \qquad \omega = \frac{d\theta}{dt} = \frac{\pi^2}{12}\sin\left(\frac{\pi t}{2}\right) \qquad \alpha = \frac{d^2\theta}{dt^2} = \frac{\pi^3}{24}\cos\left(\frac{\pi t}{2}\right)$$

$$[\Sigma M_F = m\bar{a}d]$$

$$mg(.6) - D\cos\theta(.6) = ma_n(.6\cos\theta - .48\sin\theta) + ma_t(.48\cos\theta + .6\sin\theta)$$

$$[\Sigma F_n = ma_n] \qquad F_n - D - mg\cos\theta = ma_n = mr\omega^2$$

$$[\Sigma F_t = ma_t] \qquad mg\sin\theta - F_t = ma_t = mr\alpha$$

D can be found from the first equation. With D known, F_n and F_t are easily obtained from the second and third equations. The magnitude of F is then found from,

$$F = \sqrt{F_n^2 + F_t^2}$$

It is hopefully obvious from the above that D and F are functions of time. The problem statement, however, asks us to plot these forces versus θ. Since θ is also known as a function of time we can obtain the required graph through parametric plotting. This will require that we know the time at which $\theta = \pi/3$. This is easily accomplished from the given expression for θ

$$\frac{\pi}{3} = \frac{\pi}{6}\left(1 - \cos\frac{\pi t}{2}\right) \qquad t = 2 \text{ seconds}$$

MATLAB Scripts

```
%%%%%%%%%%%%%%%%%%%% Script #1 %%%%%%%%%%%%%%%%%%%%%%%%%%%
% This script solves the first equation in the problem
% formulation section for D.
syms m g th an at D
eqn = .6*m*g-.6*D*cos(th)-m*an*(.6*cos(th)-.48*sin(th))-
m*at*(.48*cos(th)+.6*sin(th));
D = solve(eqn,D)
% The above solves the equation "eqn = 0". Be sure to
% remember to re-write all your equations in this form
%%%%%%%%%%%%%%% end of script %%%%%%%%%%%%%%%%%%%%%%%%%%%
```

Output of script #1

D =

1/5*m*(5*g-5*an*cos(th)+4*an*sin(th)-4*at*cos(th)-5*at*sin(th))/cos(th)

This result is used to plot D and F versus θ in the following script.

```
%%%%%%%%%%%%%%%%%%%% Script #1 %%%%%%%%%%%%%%%%%%%%%%%%%%%
% This script plots D and F as functions of theta.
m = 200; g = 9.81; r = 1.2;
```

```
t = 0:0.01:2;
th = pi/6*(1-cos(pi*t/2));
omega = 1/12*pi^2*sin(1/2*pi*t);
alpha = 1/24*pi^3*cos(1/2*pi*t);
an = 1.2*omega.^2;
at = 1.2*alpha;
D = 1/5*m*(5*g-5*an.*cos(th)+4*an.*sin(th)-4*at.*cos(th)-
5*at.*sin(th))./cos(th);
Fn = m*an+m*g*cos(th)+D;
Ft = m*g*sin(th)-m*at;
F = sqrt(Fn.^2+Ft.^2);
plot(th*180/pi, D/1000, th*180/pi, F/1000)
axis([0 60 0 6])
xlabel('theta (deg)')
title('Force (kN)')
%%%%%%%%%%%%%% end of script %%%%%%%%%%%%%%%%%%%%%%%%%%%
```

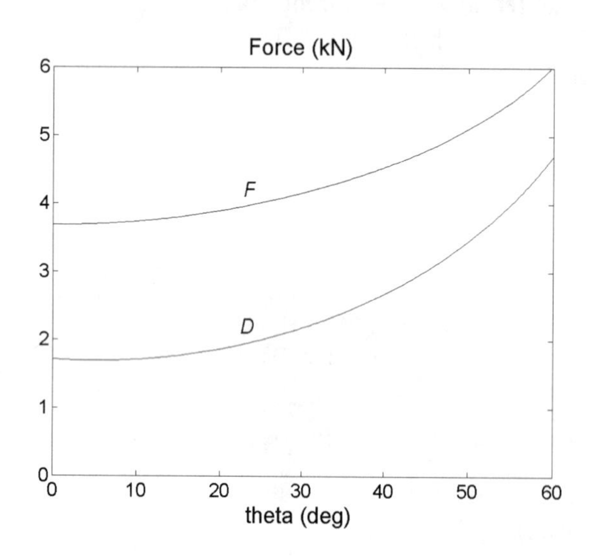

6.2 Sample Problem 6/4 (Fixed-Axis Rotation)

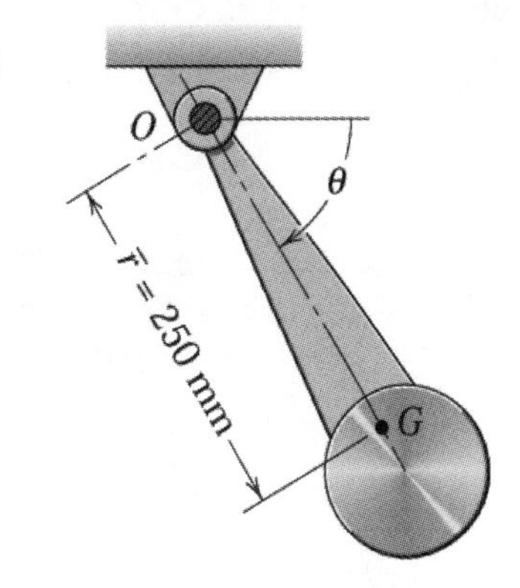

The pendulum has a mass of 7.5 kg with a mass center at G and a radius of gyration about the pivot O of 295 mm. If the pendulum is released from rest when $\theta = 0$, plot the total force supported by the bearing at O along with its normal and tangential components as a function of θ. Let θ range between 0 and 180°.

Problem Formulation

The free body and mass acceleration diagrams are identical to those in the sample problem. The main difference in approach is that we will obtain results at an arbitrary angle θ rather than at 60°.

$$[\Sigma M_O = I_O \alpha] \qquad m g \bar{r} \cos\theta = m k_0^2 \alpha \qquad \alpha = \frac{g\bar{r}}{k_0^2}\cos\theta$$

$$[\omega d\omega = \alpha d\theta] \qquad \int_0^{\omega} \omega d\omega = \frac{g\bar{r}}{k_0^2}\int_0^{\theta}\cos\theta d\theta = \frac{g\bar{r}}{k_0^2}\sin\theta$$

$$\omega^2 = \frac{2g\bar{r}}{k_0^2}\sin\theta$$

$$[\Sigma F_n = m\bar{r}\omega^2] \quad O_n - mg\sin\theta = m\bar{r}\omega^2$$

$$[\Sigma F_t = m\bar{r}\alpha] \quad -O_t + mg\cos\theta = m\bar{r}\alpha$$

After substituting for α and ω we have,

$$O_n = mg\left(1 + 2\frac{\bar{r}^2}{k_0^2}\right)\sin\theta \qquad\qquad O_t = mg\left(1 - \frac{\bar{r}^2}{k_0^2}\right)\cos\theta$$

The magnitude of the force at O is,

$$O = \sqrt{(O_n)^2 + (O_t)^2}$$

After substituting $\bar{r} = 0.25$ m, $k_0 = 0.295$ m, m = 7.5 kg, and g = 9.81 m/s^2 all forces will be functions of θ only.

MATLAB Script

```
%%%%%%%%%%%%%%% Script %%%%%%%%%%%%%%%%%%
% This script plots the force at O and its
% components as functions of theta
rb = 0.25;
k0 = 0.295;
mg = 7.5*9.81;
theta = 0:0.01:pi;
On = mg*(1+2*(rb/k0)^2)*sin(theta);
Ot = mg*(1-(rb/k0)^2)*cos(theta);
O=sqrt(On.^2+Ot.^2);
td=180*theta/pi; % converts to degrees
plot(td,O,td,On,td,Ot)
xlabel('theta (degs)')
title('Force at O (Newtons)')
```

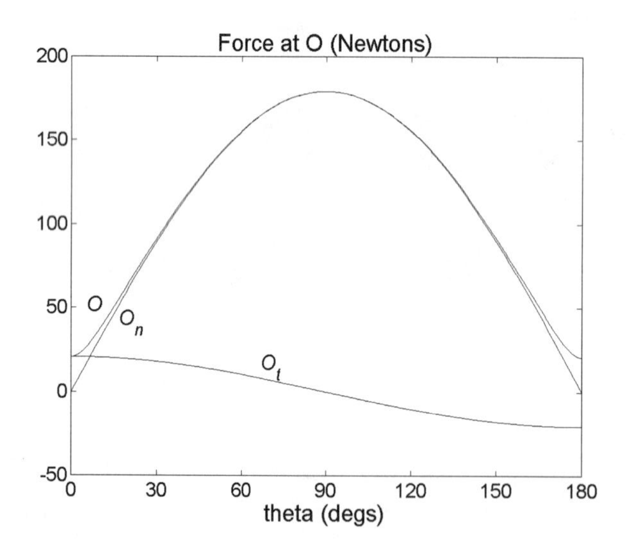

6.3 Problem 6/94 (General Plane Motion)

The slender rod of mass m and length l is released from rest in the vertical position with the small roller at end A resting on the incline. (a) Determine the initial acceleration of A (a_A) and plot a_A versus θ for $0 \leq \theta \leq 90°$. (b) Determine the maximum value of a_A over this range and the angle θ at which it occurs.

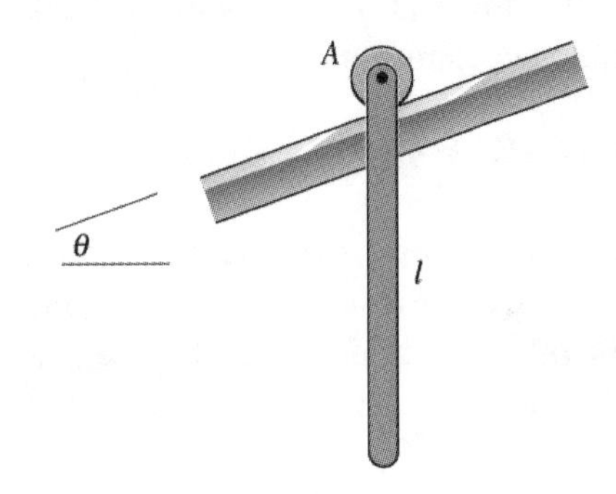

Problem Formulation

The free-body and mass acceleration diagrams are shown to the right. Shown on the mass acceleration diagram are the two components of the acceleration of the center of mass G obtained from the following kinematic relation,

$$\mathbf{a}_G = \mathbf{a}_A + (\mathbf{a}_{G/A})_n + (\mathbf{a}_{G/A})_t$$

where $(a_{G/A})_n = \frac{l}{2}\omega^2 = 0$, $(a_{G/A})_t = \frac{l}{2}\alpha$.

$$\left[\Sigma M_A = \bar{I}\alpha + m\bar{a}d\right] \qquad 0 = \frac{1}{12}ml^2\alpha + m\frac{l}{2}\alpha\frac{l}{2} - ma_A\frac{l}{2}\cos\theta$$

$$\left[\Sigma F_x = m\bar{a}_x\right] \qquad mg\sin\theta = m\left(a_A - \frac{l}{2}\alpha\cos\theta\right)$$

These two equations can be solved simultaneously to give,

$$\alpha = \frac{6(g/l)\sin\theta\cos\theta}{4 - 3\cos^2\theta} \qquad a_A = \frac{4g\sin\theta}{4 - 3\cos^2\theta}$$

At first, the answer to part (b) seems obvious. Intuitively, we would like to say that the maximum acceleration is $a_A = g$ and occurs at $\theta = 90°$. But this intuition neglects the effects of the bar's rotation upon the acceleration. As we will see below, the maximum acceleration is somewhat larger than g.

The maximum acceleration is obtained in the usual manner. The orientation where the maximum occurs is first found by solving the equation $da_A / d\theta = 0$

for θ. This angle is then substituted back into the expression for a_A to yield the maximum acceleration.

MATLAB Worksheet and Script

When using the *solve* command to solve simultaneous equations, be sure that the variables solved for are single characters. Thus, in the following we let x = a_A and y = α. Also remember that the equations solved must be in the form where the right hand side is zero. The equals sign is omitted.

```
EDU» syms x y L theta g m
EDU» eqn1 = 1/12*m*L^2*y+m*L/2*y*L/2-m*x*L/2*cos(theta)
eqn1 = 1/3*m*L^2*y-1/2*m*x*L*cos(theta)

EDU» eqn2 = m*g*sin(theta)-m*(x-L/2*y*cos(theta))
eqn2 = m*g*sin(theta)-m*(x-1/2*L*y*cos(theta))

EDU» [x,y]=solve(eqn1,eqn2)
x = -4*g*sin(theta)/(-4+3*cos(theta)^2)

y = -6*cos(theta)*g*sin(theta)/L/(-4+3*cos(theta)^2)
```

The result for x will be used in the script below to plot a_A versus θ. It is convenient to go ahead and do part (b) first as we have everything already set up.

```
EDU» aA = subs(x,g,9.81)
aA = -981/25*sin(theta)/(-4+3*cos(theta)^2)

EDU» daA = diff(aA,theta)
daA =
-981/25*cos(theta)/(-4+3*cos(theta)^2)-5886/25*sin(theta)^2/
(-4+3*cos(theta)^2)^2*cos(theta)

EDU» solve(daA,theta)
ans =
[                    1/2*pi]
[   atan(1/6*3^(1/2)*6^(1/2))]
[  -atan(1/6*3^(1/2)*6^(1/2))]
[ -atan(1/6*3^(1/2)*6^(1/2))+pi]
[  atan(1/6*3^(1/2)*6^(1/2))-pi]
```

MATLAB has found five solutions. Now we use the *eval* command to put these in a form easier to understand.

EDU» eval(ans)
ans =
   1.5708
   0.6155
  -0.6155
   2.5261
  -2.5261

Only two of the five solutions are in the range from 0 to 90°. These two solutions are π/2 (90°) and 0.6155 rad (35.3°). Substitution will reveal which is the maximum. Of course, we could also look at the plot below to see that the second solution corresponds to a maximum.

EDU» subs(aA,theta,.6155)
ans = 11.3276

EDU» subs(aA,theta,pi/2)
ans = 9.8100   % this result shouldn't be surprising

Thus, $(a_A)_{\max} = 11.33$ m/s^2 when $\theta = 35.3°$.

```
%%%%%%%%%%%%%%%%% Script %%%%%%%%%%%%%%%%
% This script plots aA versus theta
g = 9.81;
theta = 0:0.01:pi/2;
aA = -4*g*sin(theta)./(-4+3*cos(theta).^2);
plot(theta*180/pi, aA)
xlabel('theta (deg)')
ylabel('a_A (m/s^2)')
%%%%%%%%%%%%%%% end of script %%%%%%%%%%%
```

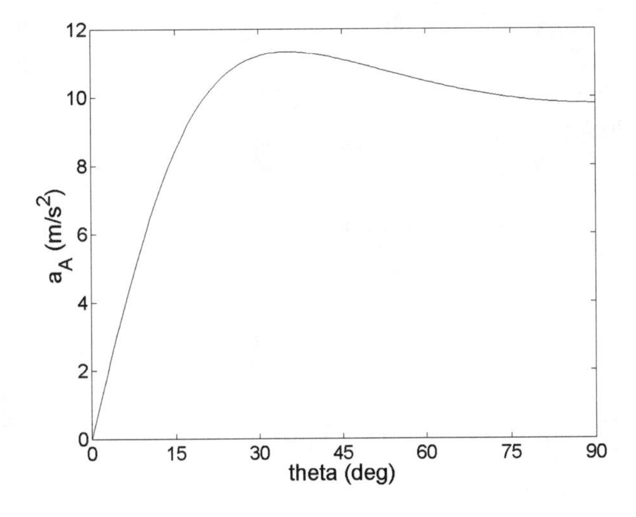

6.4 Problem 6/99 (General Plane Motion)

The uniform 12-ft pole is hinged to the truck bed and released from the vertical position as the truck starts from rest with an acceleration a. If the acceleration remains constant during the motion of the pole, derive an expression for the angular velocity ω in terms of a, g, and L where L is the length of the pole. Plot ω versus θ for $a = 3$ ft/s^2.

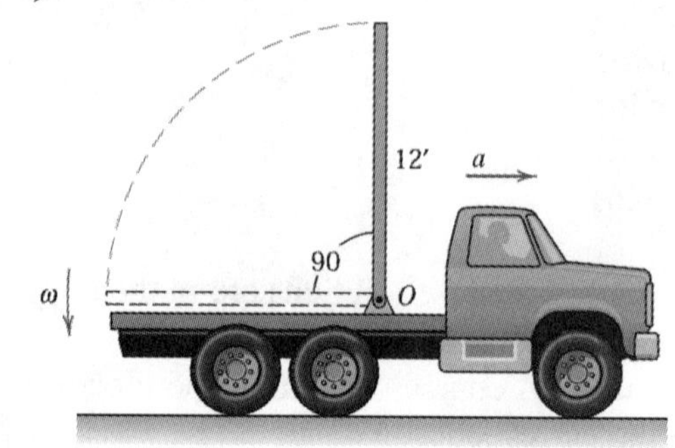

Problem Formulation

The free-body and mass acceleration diagrams are shown to the right. Shown on the mass acceleration diagram are the three components of the acceleration of the center of mass G obtained from the following kinematic relation,

FBD MAD

$$\mathbf{a}_G = \mathbf{a}_O + (\mathbf{a}_{G/O})_n + (\mathbf{a}_{G/O})_t$$

where $a_O = a$, $(a_{G/O})_n = \bar{r}\omega^2$, $(a_{G/O})_t = \bar{r}\alpha$, and $\bar{r} = L/2 = 6$ feet.

$$\left[\Sigma M_O = \bar{I}\alpha + m\bar{a}d\right] \qquad mg\frac{L}{2}\sin\theta = \frac{1}{12}mL^2\alpha + m\frac{L}{2}\alpha\frac{L}{2} - ma\frac{L}{2}\cos\theta$$

$$\alpha = \frac{3}{2L}(g\sin\theta + a\cos\theta)$$

Now we integrate the relation $\omega d\omega = \alpha d\theta$ to obtain,

$$\frac{1}{2}\omega^2 = \frac{3}{2L}\int_0^\theta (g\sin\theta + a\cos\theta)d\theta = \frac{3}{2L}(g(1-\cos\theta) + a\sin\theta)$$

$$\omega = \sqrt{\frac{3}{L}(g(1-\cos\theta) + a\sin\theta)}$$

MATLAB Script

```
%%%%%%%%%%%%%%%%%%%% Script %%%%%%%%%%%%%%%%%%%
L = 12; g = 32.2; a = 3;
theta = 0:0.01:pi/2;
omega = sqrt(3/L*(g*(1-cos(theta))+a*sin(theta)));
plot(theta*180/pi,omega)
xlabel('theta (deg)')
ylabel('omega (rad/s)')
%%%%%%%%%%%%%%%%%end of script %%%%%%%%%%%%%%%%
```

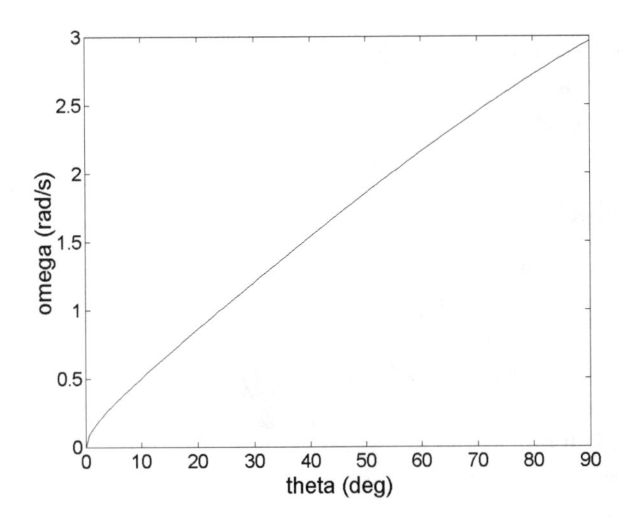

6.5 Sample Problem 6/10 (Work and Energy)

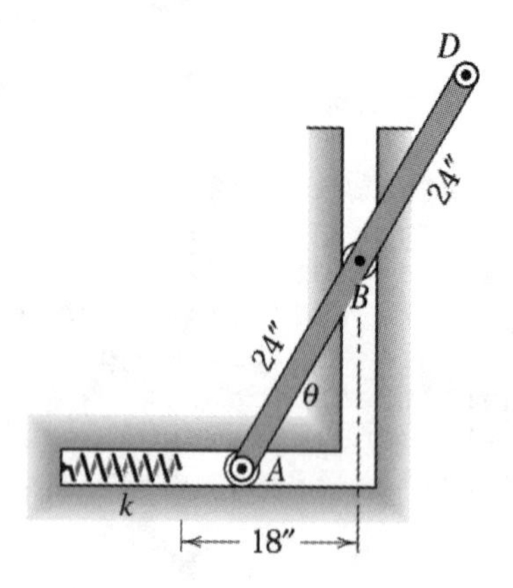

The 4-ft slender bar weighs 40 lb with a mass center at B and is released from rest in the position for which θ is essentially zero. Point B is confined to move in the smooth vertical guide, while end A moves in the smooth horizontal guide and compresses the spring as the bar falls. Plot the angular velocity of the bar and the velocities of A and B as a function of θ from 0 to 90°. The stiffness of the spring is 30 lb/in.

Problem Formulation

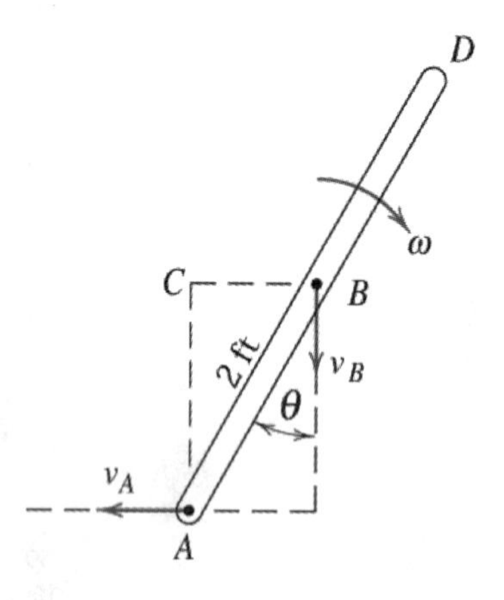

From the figure to the right, the lengths CB and CA are $2\sin\theta$ and $2\cos\theta$ respectively. Using the instantaneous center C we can write the two velocities in terms of the angular velocity ω.

$$v_A = CA\omega = 2\omega\cos\theta \qquad v_B = CB\omega = 2\omega\sin\theta$$

Now we need to divide the range for θ into two distinct intervals depending upon whether or not the spring has been engaged. Since the velocities for A and B are known in terms of ω and θ, we need to find only the angular velocity in these two intervals. From the diagram we see that A will first contact the spring at an angle $\theta = \sin^{-1}(18/24) = 0.8481$ rads (48.6°).

(a) Before the spring is engaged ($\theta \le 48.6°$).

$$[T = \frac{1}{2}m\bar{v}^2 + \frac{1}{2}\bar{I}\omega^2] \quad \Delta T = \frac{1}{2}\frac{40}{32.2}(2\omega\cos\theta)^2 + \frac{1}{2}\left(\frac{1}{12}\frac{40}{32.2}4^2\right)\omega^2$$

$$\Delta T = 0.8282\left(4 - 3\cos^2\theta\right)\omega^2$$

$$[\Delta V_g = W\Delta h] \qquad \Delta V_g = 40(2\cos\theta - 2) = 80(\cos\theta - 1)$$

We now substitute into the energy equation $U'_{1-2} = \Delta T + \Delta V_g = 0$,

$0 = 0.8282\left(4 - 3\cos^2\theta\right)\omega^2 + 80(\cos\theta - 1)$, from which we find

$$\omega = 9.829\sqrt{\frac{1-\cos\theta}{4-3\cos^2\theta}}$$

(b) After the spring is engaged ($48.6 \le \theta \le 90°$). The kinetic and potential energies are the same as in part (a). At any angle θ, point A has moved $2\sin\theta$ feet to the left. Thus, the spring is compressed by $2\sin\theta - 18/12$ feet.

$$[V_e = \frac{1}{2}kx^2] \qquad \Delta V_e = \frac{1}{2}\left(30\frac{lb}{in}\right)\left(12\frac{in}{ft}\right)\left(2\sin\theta - \frac{18}{12}\right)^2 - 0$$

$$\Delta V_e = 180\left(2\sin\theta - \frac{3}{2}\right)^2$$

Again, we substitute into the energy equation $U'_{1-2} = \Delta T + \Delta V_g + \Delta V_e = 0$,

$$0 = 0.8282\left(4-3\cos^2\theta\right)\omega^2 + 80(\cos\theta - 1) + 180\left(2\sin\theta - \frac{3}{2}\right)^2$$

$$\omega = 2.457\sqrt{\frac{216\sin\theta - 16\cos\theta - 144\sin^2\theta - 65}{4-3\cos^2\theta}}$$

MATLAB Scripts

In the following, terms ending with "_a" are for part (a) where θ is between 0 and 48.6° while terms ending with "_b" are for part (b) where θ is greater than 48.6°.

```
%%%%%%%%%%%%%%% Script #1 %%%%%%%%%%%%%%%%
% This script solves the energy equation for omega
% for cases a and b
syms theta omega
DT = 0.8282*(4-3*cos(theta)^2)*omega^2;
DVg = 80*(cos(theta)-1);
DVe = 180*(2*sin(theta)-3/2)^2;
U12_a = DT + DVg
U12_b = DT + DVg + DVe
omega_a = solve(U12_a,omega)
omega_b = solve(U12_b,omega)
%%%%%%%%%%%%%%%%% end of script %%%%%%%%%%%%%%
```

Output of script #1

U12_a =
(4141/1250-12423/5000*cos(theta)^2)*omega^2+80*cos(theta)-80

U12_b =
(4141/1250-12423/5000*cos(theta)^2)*omega^2+80*cos(theta)-80+180*(2*sin(theta)-3/2)^2

omega_a =
[200/(-16564+12423*cos(theta)^2)*41410^(1/2)*((-4+3*cos(theta)^2)*(cos(theta)-1))^(1/2)]

[-200/(-16564+12423*cos(theta)^2)*41410^(1/2)*((-4+3*cos(theta)^2)*(cos(theta)-1))^(1/2)]

It shouldn't be surprising that MATLAB has found two solutions to the equation. As you can see, they have the same magnitude with one being positive and the other negative. The solution of interest is the positive one which turns out to be the second solution. This solution is copied and pasted into script #2.

omega_b =
[50/(4141+12423*sin(theta)^2)*(-(41410+124230*sin(theta)^2)*(16*cos(theta)+65+144*sin(theta)^2-216*sin(theta)))^(1/2)]

[-50/(4141+12423*sin(theta)^2)*(-(41410+124230*sin(theta)^2)*(16* cos(theta)+65+144*sin(theta)^2-216*sin(theta)))^(1/2)]

In this case, the first solution is positive.

```
%%%%%%%%%%%%%%% Script #2 %%%%%%%%%%%%%%%%
% This script plots omega versus theta. Note that
% two range variables are used to divide the range
% for theta into the two regions.
% a = theta for theta < 48.6 deg. (.8481 rads)
% b = theta for theta > 48.6 deg.
% The results for omega are copied and pasted from
% the output of script #1
a = 0:0.001:0.8481;
b = 0.8481:0.001:pi/2;
omega_a = -200./(-16564+12423*cos(a).^2)*41410^(1/2).*((-
4+3*cos(a).^2).*(cos(a)-1)).^(1/2);
omega_b = 50./(4141+12423*sin(b).^2).*(-
(41410+124230*sin(b).^2).*(16*cos(b)+65+144*sin(b).^2-
216*sin(b))).^(1/2);
plot(a*180/pi, omega_a, b*180/pi, omega_b)
```

```
xlabel('theta (degrees)')
ylabel('omega (rad/sec)')
%%%%%%%%%%%%%%% end of script %%%%%%%%%%%%%%%
```

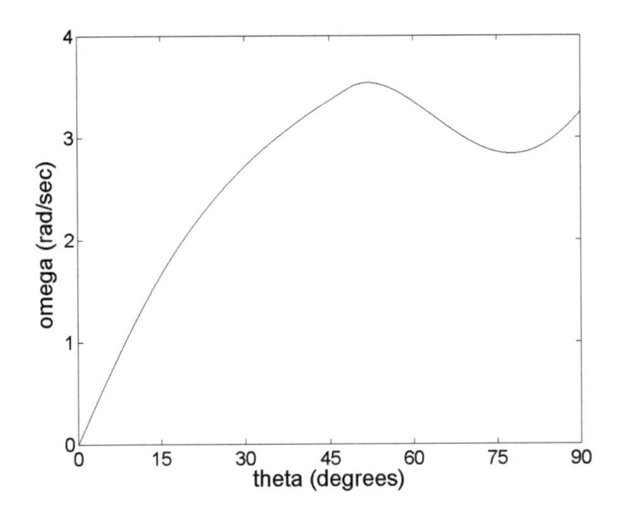


```
%%%%%%%%%%%%%%% Script #3 %%%%%%%%%%%%%%%
% This script plots vA and vB versus theta.
a = 0:0.001:0.8481;
b = 0.8481:0.001:pi/2;
omega_a = -200./(-16564+12423*cos(a).^2)*41410^(1/2).*((-
4+3*cos(a).^2).*(cos(a)-1)).^(1/2);
omega_b = 50./(4141+12423*sin(b).^2).*(-
(41410+124230*sin(b).^2).*(16*cos(b)+65+144*sin(b).^2-
216*sin(b))).^(1/2);
vA_a = 2*omega_a.*cos(a);
vA_b = 2*omega_b.*cos(b);
vB_a = 2*omega_a.*sin(a);
vB_b = 2*omega_b.*sin(b);
plot(a*180/pi, vA_a, b*180/pi, vA_b, a*180/pi, vB_a, b*180/pi, vB_b)
xlabel('theta (degrees)')
ylabel('velocity (ft/sec)')
%%%%%%%%%%%%%%% end of script %%%%%%%%%%%%%%%
```

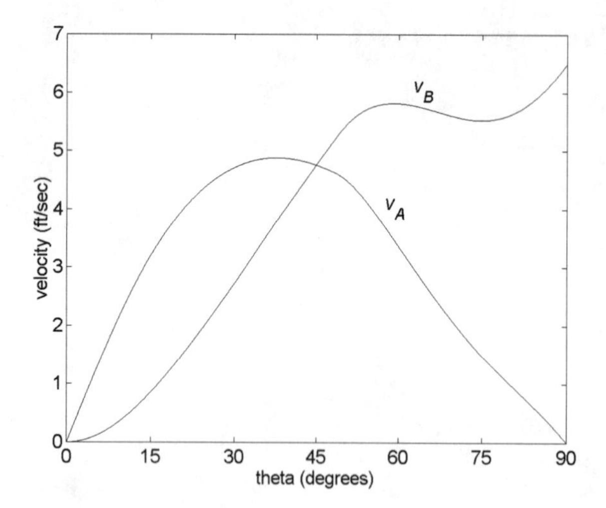

For Further Study

A striking feature of the velocity curves above is the sudden change in shape when the spring is engaged at about $\theta = 49°$. The details depend very much upon the relative magnitudes of the weight and spring constant. To illustrate, the figure below shows the angular velocity for several different values of the spring constant k.

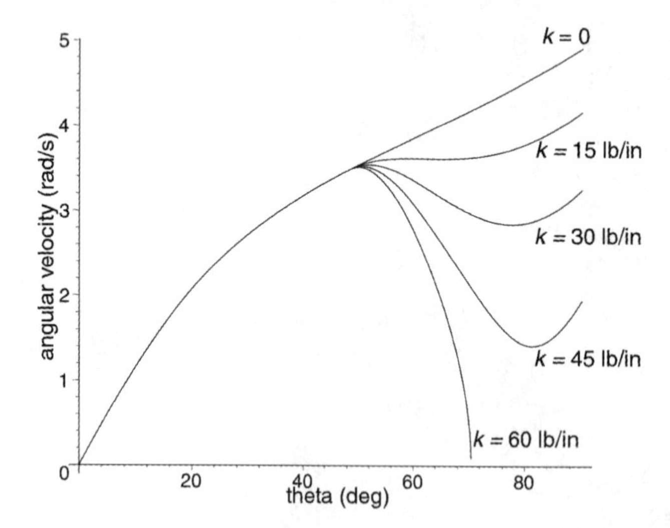

Note that for stiff springs, the angular velocity goes to zero before reaching $\theta = 90°$. The physical explanation for this is that, for a stiff spring, the bar will rebound before it reaches the horizontal position.

6.6 Problem 6/204 (Impulse/Momentum)

Determine the minimum velocity v that the wheel may have and just roll over the obstruction. The centroidal radius of gyration of the wheel is k, and it is assumed that the wheel does not slip. Plot v versus h for three cases: k = ½, ¾, and 1 m. For each case take $r = 1$ m.

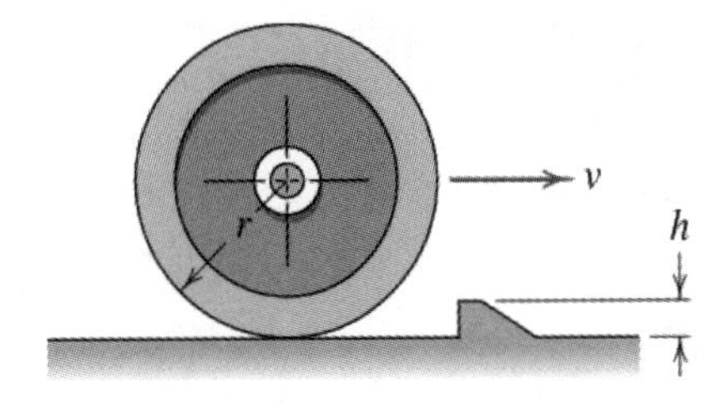

Problem Formulation

During Impact: Conservation of Angular Momentum

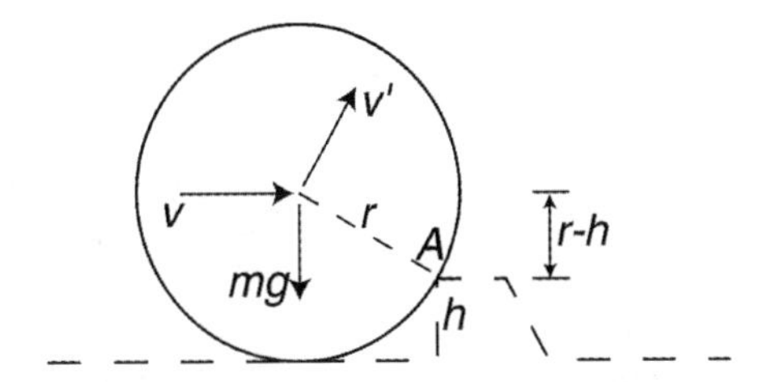

As usual, we neglect the angular impulse of the weight during the short interval of impact. With this assumption we have conservation of angular momentum about point A. Immediately before impact, the center of the wheel is not moving in a circular path about A and we need to use the formula for general plane motion. Note that $\bar{I} = mk^2$.

$$H_A = \bar{I}\omega + m\bar{v}d = mk^2\omega + mv(r-h) = mk^2\frac{v}{r} + mv(r-h)$$

We will use primes to denote the state immediately after impact. Since the wheel now rotates about A we can use the simpler formula $H_A = I_A\omega$. Note that, by the parallel axis theorem, $I_A = \bar{I} + mr^2 = m(k^2 + r^2)$.

$$H_A' = I_A\omega' = (k^2 + r^2)\frac{v'}{r}$$

Setting $H_A = H_A'$ and solving yields,

$$v' = v\left(1 - \frac{rh}{k^2 + r^2}\right)$$

After Impact: Work-Energy

$$\Delta T + \Delta V_g = 0 = \frac{1}{2}I_A\left(0^2 - \omega'^2\right) + mgh$$

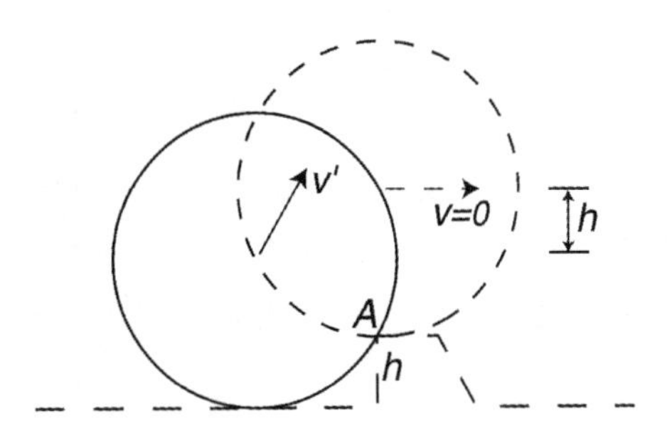

$$\frac{1}{2}m\left(k^2+r^2\right)\left(\frac{v'}{r}\right)^2=mgh$$

Substituting the result for v' into the above equation followed by simplification yields,

$$v=\frac{r\sqrt{2gh\left(k^2+r^2\right)}}{k^2+r^2-rh}$$

MATLAB Scripts

```
%%%%%%%%%%%%%% Script #1 %%%%%%%%%
% symbolic solution for v
syms m g k r h v vp
Ib = m*k^2; IA = m*(k^2+r^2);
HA = Ib*v/r+m*v*(r-h);
HAp = IA*vp/r;
vp = solve(HA-HAp,vp) % conservation of angular momentum
eqn = 1/2*IA*vp^2/r^2-m*g*h; % work/energy
solve(eqn,v)
%%%%%%%%%% end of script %%%%%%%%%
```

Output of script #1

vp =
v*(k^2+r^2-r*h)/(k^2+r^2)

ans =
[(2*g*h*r^2+2*g*h*k^2)^(1/2)*r/(k^2+r^2-r*h)]
[-(2*g*h*r^2+2*g*h*k^2)^(1/2)*r/(k^2+r^2-r*h)]

The first solution is the one we want.

```
%%%%%%%%%%%%%% Script #2 %%%%%%%%%
% This script plots v versus h for the specified values of k.
% The result for v below was obtained by substituting g = 9.81
% and r = 1 in to the symbolic result found in script #1
v = inline('4.429*sqrt(h+h*k^2)/(1+k^2-h)');
v = vectorize(v); % puts "."'s in expression
h = 0:0.01:1;
plot(h,v(h,1/2),h,v(h,3/4),h,v(h,1))
xlabel('h (m)')
ylabel('v (m/s)')
%%%%%%%%%% end of script %%%%%%%%%
```

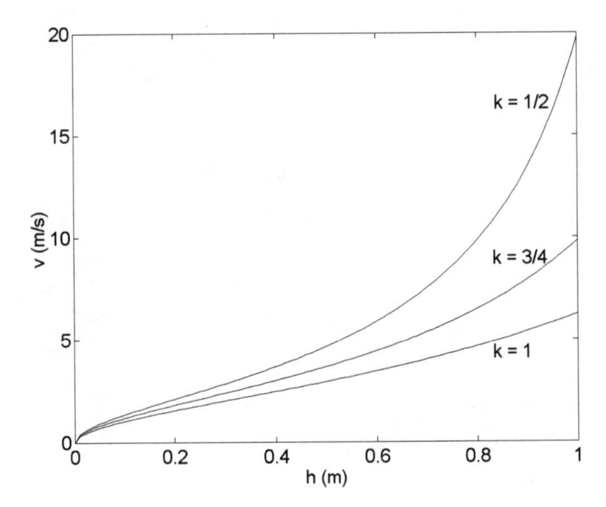

20
15
10
5
0
v (m/s)
0
0.2
0.4
0.6
0.8
1
h (m)
k = 1/2
k = 3/4
k = 1

INTRODUCTION TO THREE-DIMENSIONAL DYNAMICS OF RIGID BODIES

7

This chapter presents a brief introduction to rigid body dynamics in three dimensions. Problem 7.1 considers a solid right-circular cone that rolls on a flat surface in such a manner that the cone rotates about a fixed point. The effects of the geometry of the cone upon its angular acceleration are studied. In problem 7.2, the general 3-D motion of three connected bars is investigated. In particular, the angular velocities of two of the bars are plotted versus the length of the third bar. *solve* is used to solve four equations symbolically for four unknowns. In problem 7.3 we consider a bent plate rotating about a fixed axis. The problem illustrates a simplified version of what engineers might do in a real design situation. Two dimensions of the bent plate are left as variables and the objective of the problem is to find all suitable values of those dimensions which satisfy several constraints simultaneously. Here we illustrate a graphical approach to this type of design problem.

7.1 Problem 7/27 (Rotation about a Fixed Point)

The solid right-circular cone of base radius r and height h rolls on a flat surface without slipping. The center B of the circular base moves in a circular path around the z-axis with a constant speed v. Find the magnitude of the angular acceleration α and plot α versus h for $r = 0.5$ m and $v = 1$ m/s. Let h range between 1 and 5 meters.

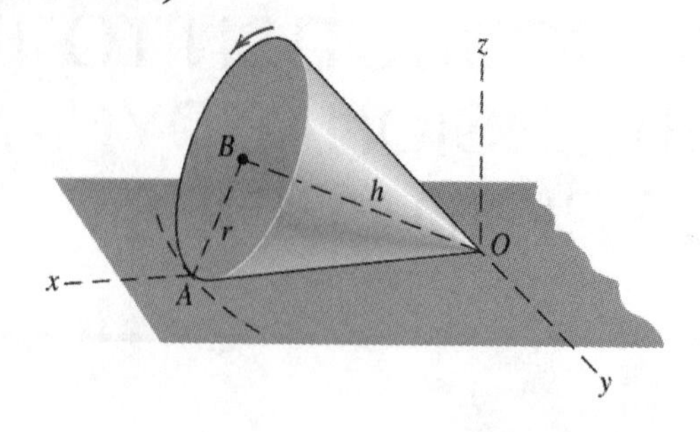

Problem Formulation

From the diagram we first note that $v = (r\cos\gamma)\omega$ so that

$$\boldsymbol{\omega} = \frac{v}{r\cos\gamma}\mathbf{i}$$

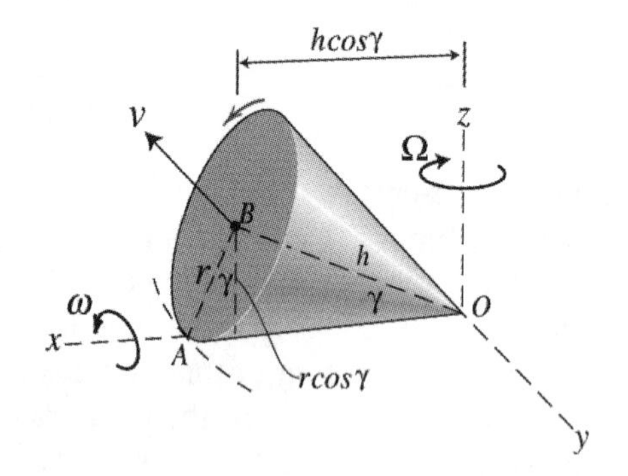

Since $\boldsymbol{\omega}$ is constant we have (see equation 7/3 in your text),

$$\boldsymbol{\alpha} = \boldsymbol{\Omega} \times \boldsymbol{\omega}$$

Since $v = (h\cos\gamma)\Omega$ (see figure),

$$\boldsymbol{\Omega} = -\frac{v}{h\cos\gamma}\mathbf{k}$$

$$\boldsymbol{\alpha} = -\frac{v^2}{rh\cos^2\gamma}\mathbf{k}\times\mathbf{i} = -\frac{v^2}{rh\cos^2\gamma}\mathbf{j}$$

$$\alpha = \frac{v^2}{rh\cos^2\gamma}$$

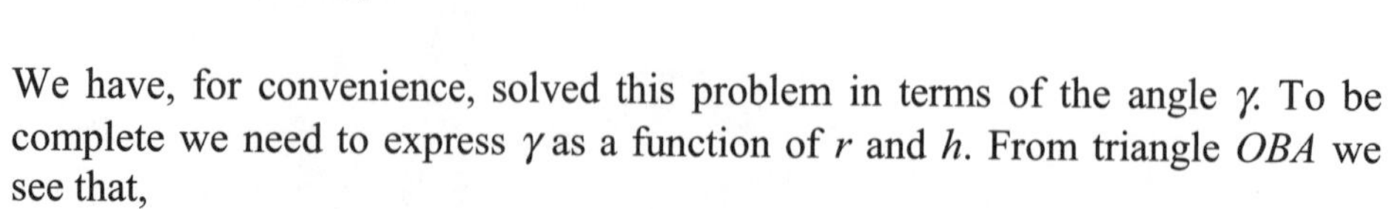

We have, for convenience, solved this problem in terms of the angle γ. To be complete we need to express γ as a function of r and h. From triangle OBA we see that,

$$\gamma = \tan^{-1}(r/h)$$

MATLAB Script

```
%%%%%%%%%%%%%%%%%%%%%% Script %%%%%%%%%%%%%%%
v = 1; r = 0.5;
h = 1:0.05:5;
gamma = atan(r./h);
alpha = v^2./h./r./cos(gamma).^2;
plot (h, alpha)
xlabel('h (m)')
ylabel('alpha (rad/s^2)')
%%%%%%%%%%%%%%%%%%end of script %%%%%%%%%%%%
```

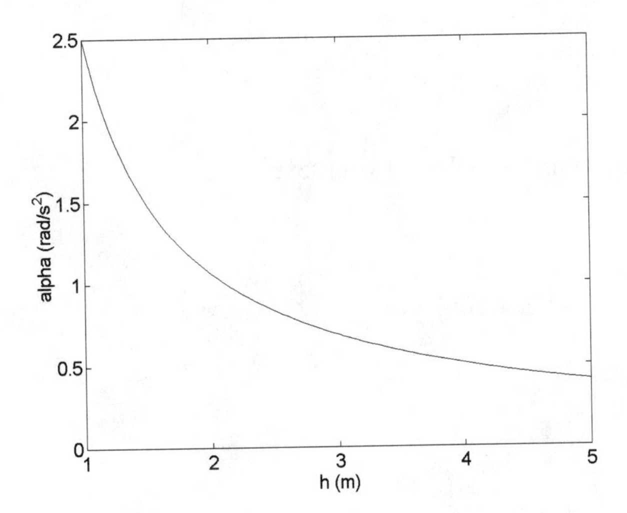

7.2 *Sample Problem 7/3 (General Motion)*

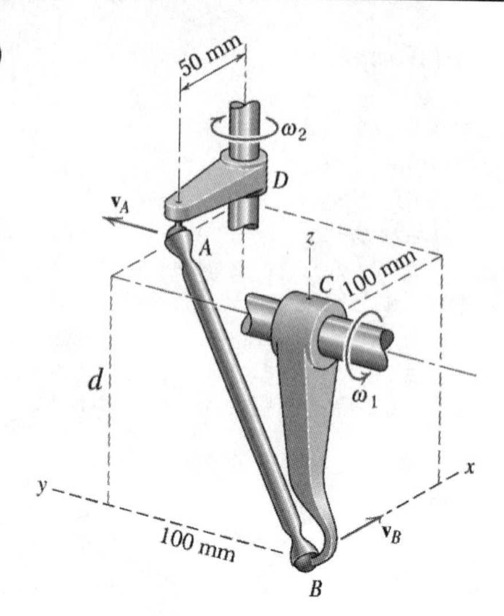

Crank *CB* rotates about the horizontal axis with an angular velocity ω_1 = 6 rad/s, which is constant for a short interval of motion that includes the position shown. Link *AB* has a ball-and-socket fitting on each end and connects crank *DA* with *CB*. Let the length of crank *CB* be *d* mm (instead of 100 mm as in the sample problem in your text) and plot ω_2 and ω_n as a function of *d* for $0 \le d \le 200$ mm. ω_2 is the angular velocity of crank *DA* while ω_n is the angular velocity of link *AB*.

Problem Formulation

Our analysis will follow closely that in the sample problem in your text.

$$\mathbf{v}_A = \mathbf{v}_B + \boldsymbol{\omega}_n \times \mathbf{r}_{A/B}$$

where $\mathbf{v}_A = 50\omega_2\mathbf{j}$ $\qquad \mathbf{v}_B = 6d\mathbf{i}$ $\qquad \mathbf{r}_{A/B} = 50\mathbf{i} + 100\mathbf{j} + d\mathbf{k}$

Substitution into the velocity equation gives

$$50\omega_2\mathbf{j} = 6d\mathbf{i} + \begin{vmatrix} \mathbf{i} & \mathbf{j} & \mathbf{k} \\ \omega_{nx} & \omega_{ny} & \omega_{nz} \\ 50 & 100 & d \end{vmatrix}$$

Expanding the determinant and equating the **i**, **j**, and **k** components yields the following three equations

$$d(6 + \omega_{ny}) - 100\omega_{nz} = 0 \qquad 50(\omega_2 - \omega_{nz}) + d\omega_{nx} = 0 \qquad 2\omega_{nx} - \omega_{ny} = 0$$

At this point we have three equations with four unknowns. As explained in the sample problem in your text, the fourth equation comes by requiring $\boldsymbol{\omega}_n$ to be normal to $\mathbf{v}_{A/B}$

$$\boldsymbol{\omega}_n \bullet \mathbf{r}_{A/B} = 50\omega_{nx} + 100\omega_{ny} + d\omega_{nz} = 0$$

These four equations will be solved simultaneously for ω_2, ω_{nx}, ω_{ny}, *and* ω_{nz}. Once this is done,

$$\omega_n = \sqrt{\omega_{nx}^2 + \omega_{ny}^2 + \omega_{nz}^2}$$

MATLAB Scripts

```
%%%%%%%%%%%%%%%% Script #1 %%%%%%%%%%%%%%%%%%
% This script solves four equations for
% omega_2 (w) and the three components of
% omega_n (x, y, and z). The magnitude of
% omega_n is then found from its components.
syms w x y z d
eqn1 = d*(6+y)-100*z;
eqn2 = 50*(w-z)+d*x;
eqn3 = 2*x-y;
eqn4 = 50*x+100*y+d*z;
[w,x,y,z]=solve(eqn1,eqn2,eqn3,eqn4)
omega_n = sqrt(x^2+y^2+z^2);
omega_n = simplify(omega_n)
%%%%%%%%%%%%%% end of script %%%%%%%%%%%%%%%%
```

Output of script #1

w = 3/50*d (ω_2)

x = -3*d^2/(d^2+12500)

y = -6*d^2/(d^2+12500)

z = 750*d/(d^2+12500)

omega_n = 3*5^(1/2)*(d^2/(d^2+12500))^(1/2) (ω_n)

```
%%%%%%%%%%%%%%%% Script #2 %%%%%%%%%%%%%%%%%%
% This script plots omega_2 and the magnitude of
% omega_n as functions of d
d = 0:0.5:200;
omega_2 = 3/50*d;
omega_n = 3*5^(1/2)*(d.^2./(d.^2+12500)).^(1/2);
plot(d, omega_2, d, omega_n)
xlabel('d (mm)')
title('angular velocity (rad/s)')
%%%%%%%%%%%%%% end of script %%%%%%%%%%%%%%%%
```

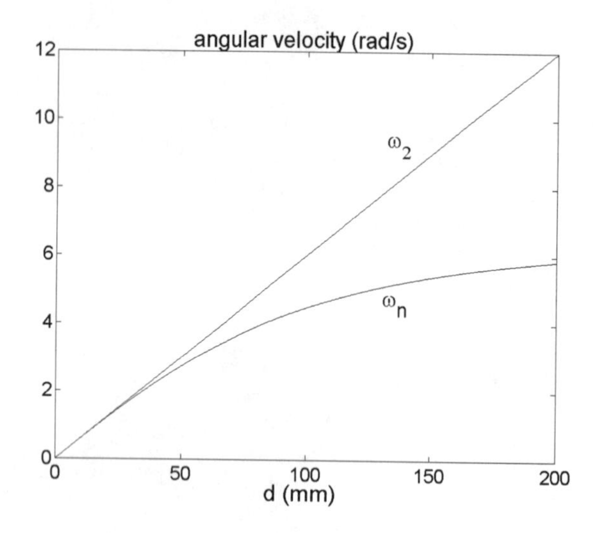

7.3 Sample Problem 7/6 (Kinetic Energy)

The bent plate has a mass of 70 kg per square meter of surface area and revolves around the z-axis at the rate $\omega = 30$ rad/s. Let the dimensions of part B be a and b where a is the dimension parallel to the x-axis and b is the dimension parallel to the z-axis. Part A remains unchanged. (a) Find all suitable values for a and b which satisfy the following conditions: $a \leq 0.2$ m, $b \leq 0.6$ m, and $15 \leq T \leq 30$ J where T is the kinetic energy of the plate. (b) Find a and b for the case where $T = 40$ J and $H_O = 5$ N•m•s where H_O is the magnitude of the angular momentum about O.

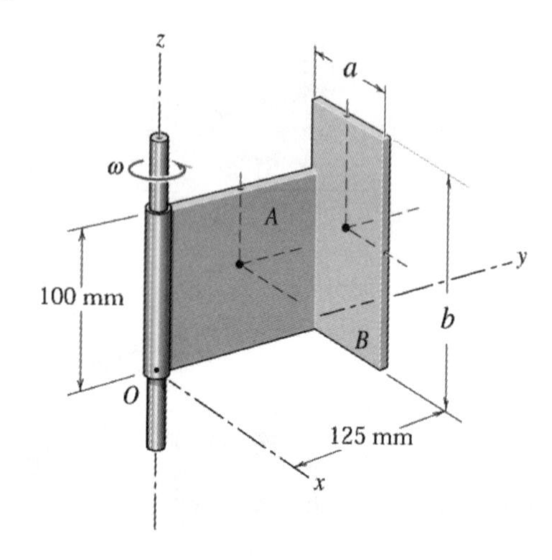

Problem Formulation

Substitution of $\omega_x = 0$, $\omega_y = 0$, and $\omega_z = \omega$ into equations 7/11 and 7/18 of your text yields.

$$\mathbf{H}_O = \omega(-I_{xz}\mathbf{i} - I_{yz}\mathbf{j} + I_{zz}\mathbf{k})$$

$$H_O = \sqrt{H_{Ox}^2 + H_{Oy}^2 + H_{Oz}^2} = \omega\sqrt{I_{xz}^2 + I_{yz}^2 + I_{zz}^2}$$

$$T = \frac{1}{2} I_{zz} \omega^2$$

The moments and products of inertia for part A remain unchanged. For part B, $m_B = 70ab$ where a and b are in meters. The moments and products of inertia for part B are

$$I_{zz} = \bar{I}_{zz} + md^2 = \frac{m_B}{12} a^2 + m_B \left((.125)^2 + \left(\frac{a}{2} \right)^2 \right)$$

$$I_{xz} = \bar{I}_{xz} + md_x d_z = 0 + m_B \left(\frac{a}{2} \right) \left(\frac{b}{2} \right)$$

$$I_{yz} = \bar{I}_{yz} + md_y d_z = 0 + m_B (0.125) \left(\frac{b}{2} \right)$$

The total moment and products of inertia are found by adding the above to those found for part A (see the sample problem in your text). After substituting for m_B and simplifying we have,

$$I_{zz} = 0.00456 + 1.094ab + 23.33a^3 b$$

$$I_{xz} = 17.5a^2 b^2 \qquad I_{yz} = 0.00273 + 4.375ab^2$$

Substitution of these results will give H_O and T as functions of a and b.

Part (a)

The most efficient way to show the acceptable ranges for a and b is to find the required relationship between these two dimensions in order to satisfy the upper and lower bounds on T. This is accomplished by substituting these bounds for T in the equation above and then solving that equation for b as a function of a. To illustrate, consider the lower limit on T (15 J). Substituting $T = 15$ into the equation above gives

$$T = 15 = \frac{1}{2} I_{zz} (30)^2 = 2.052 + 492.2ab + 10{,}500a^3 b$$

Solving for b, $\quad b = \dfrac{0.078921}{a(3 + 64a^2)} \qquad$ (for $T = 15$ J)

A similar result can be obtained for the upper limit,

$$b = \frac{0.17035}{a(3 + 64a^2)} \qquad \text{(for } T = 30 \text{ J)}$$

Plotting these two functions defines the acceptable regions for a and b.

Part (b)

This is similar to (a) except that we solve two equations (T = 40 and H_0 = 5) simultaneously for two unknowns, a and b. The result is a = 0.1211 m and b = 0.4852 m.

MATLAB Scripts

```
%%%%%%%%%%%%%%% Script #1 %%%%%%%%%%%%%%%%%%%%%%%
% The first part of this script solves the equations
% T = 15 and T = 30 symbolically for b in terms of a.
% These two results are named b15 and b30 respectively.
% When using the symbolic solve it is best to rewrite
% the equations in the form expression = 0. The "=0"
% is automatically understood by MATLAB and doesn't
% have to be typed in. In the following, our two
% expressions are named eqn1 and eqn2.
syms a b
mB = 70*a*b;
omega = 30;
Izz = mB/12*a^2+mB*(0.125^2+(a/2)^2)+0.00456;
Ixz = mB*a/2*b/2;
Iyz = mB*0.125*b/2+0.00273;
H0 = omega*sqrt(Ixz^2+Iyz^2+Izz^2)
T = 1/2*Izz*omega^2
eqn1 = T - 15;
eqn2 = T - 30;
b15 = solve(eqn1,b)
b30 = solve(eqn2,b)
% This part of the script solves the two equations
% for part b.
eqn3 = T - 40;
eqn4 = H0 - 5;
[a,b]=solve(eqn3,eqn4)
%%%%%%%%%%% end of script %%%%%%%%%%%%%%%%%%%%%%%
```

Output of script #1

```
H0 =
1/10000*(27562500000000*a^4*b^4+1722656250000*a^2*b^4+2149875000*a*b^2+25
42185+49000000000000*a^6*b^2+4593750000000*a^4*b^2+19152000000*a^3*b+107
666015625*a^2*b^2+897750000*a*b)^(1/2)
```

T =2625*a^3*b+31500*a*b*(1/64+1/4*a^2)+513/250

b15 =8632/109375/a/(64*a^2+3)

b30 = 18632/109375/a/(64*a^2+3)

a =

[-.118672955347671873813795187]
[-.924961825359232183759243317e-1-.204059123637207694380947 26*i]
[-.924961825359232183759243317e-1+.204059123637207694380947 26*i]
[-.257514497287494286960690298e-2-.254917314610665797875638 83*i]
[-.257514497287494286960690298e-2+.254917314610665797875638 83*i]
[.349640543257908784471004656e-2-.265040090095767094717337313*i]
[.349640543257908784471004656e-2+.26504009009576709471733731*i]
[.903798369106820783468910094e-1-.204312206452426767071840559*i]
[.903798369106820783468910094 5e-1+.204312206452426767071840 5*i]
[.121063125678745863921655544]

b =

[-.499591682557401531621796988]
[.286519678237340239067367533+.281089760611403745879347469*i]
[.286519678237340239067367533-.281089760611403745879347469*i]
[.643757758855131608495630 80e-1-.778487038145858201006376746*i]
[.643757758855131608495630 80e-1+.778487038145858201006376 74*i]
[-.536708278185586943061676 4e-1-.579398660286892007965618348*i]
[-.536708278185586943061676 4e-1+.579398660286892007965618 34*i]
[-.298105878686806359955473 22+.284559015898182670929445 34*i]
[-.298105878686806359955473 22-.284559015898182670929445341*i]
[.485167563521434010754982656]

We see from the above that MATLAB has found ten solutions to our equations. Of these only the first and last are real. The first solution has negative values for a and b and thus can be excluded. This leaves only the last solution,

$$a = 0.1211 \text{ m};\ b = 0.4852 \text{ m}$$

```
%%%%%%%%%%%%%% Script #2 %%%%%%%%%%%%%%%%%%%%%
% This script plots the two expressions for b
% obtained with script #1
a = 0.01:0.001:0.2;
b15 = 8632/109375./a./(64*a.^2+3);
b30 = 18632/109375./a./(64*a.^2+3);
plot(a, b15, a, b30)
xlabel('a (m)')
```

```
ylabel('b (m)')
axis([0 0.2 0 0.6])
%%%%%%%%%%% end of script %%%%%%%%%%%%%%%%%%%%%%%
```

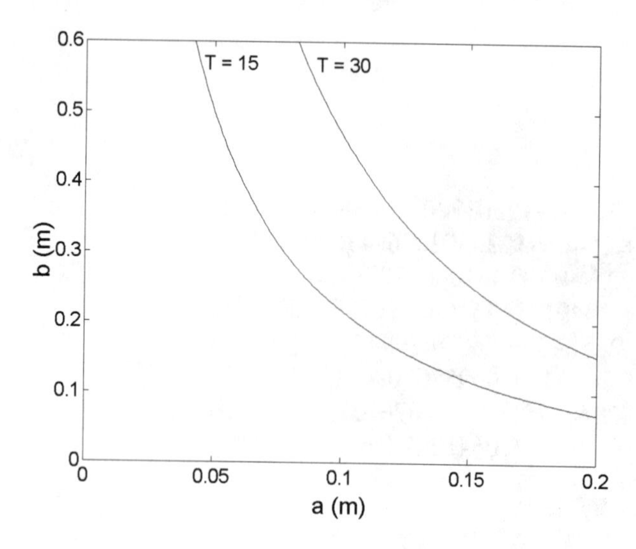

The two curves above represent the values of a and b for which T is exactly 15 or 30 J. Thus, the acceptable values of a and b satisfying the condition $15 \leq T \leq 30$ J are all those combinations lying on or between the two curves.

VIBRATION AND TIME RESPONSE 8

This chapter considers an important class of Dynamics problems that involve linear or angular oscillations of a body or structure about some equilibrium position or configuration. Very few of the homework problems in your text require you to actually plot the oscillations of a body versus time. For this reason, three of the four problems in this chapter will involve such plots. This is very useful in visualizing the time response of a vibrating system, especially for the case of damped or forced vibrations. Problem 8.1 looks at the effects of damping coefficient upon time response while Problem 8.2 considers the effects of initial conditions. In problem 8.3 we plot the magnification factor versus the frequency of a forced input and then determine the magnitude of the maximum magnification factor and the frequency at which occurs. As usual, this is accomplished using *diff* and *solve*. Problem 8.4 considers the angular oscillation of a rigid body and studies the effect of the geometry of the body on the natural frequency. Throughout this supplement our method of choice for solving equations has been to use the symbolic *solve* command. The advantage of this approach is that *solve* will seek first a symbolic solution and then, if this fails, automatically switch to a numerical solution. Occasionally this approach fails to find any solution. Problem 8.4 is an example of such a case and illustrates the alternate approach of creating a function m-file and then using *fzero* to obtain a numerical solution. This problem also takes a brief look at the errors introduced by the small angle approximations for sine, cosine and tangent.

8.1 Sample Problem 8/2 (Free Vibration of Particles)

The 8-kg body is moved 0.2 m to the right of the equilibrium position and released from rest at time t = 0. Plot the displacement as a function of time for three cases, c = 8, 32, and 56 N•s/m. The spring stiffness k is 32 N/m.

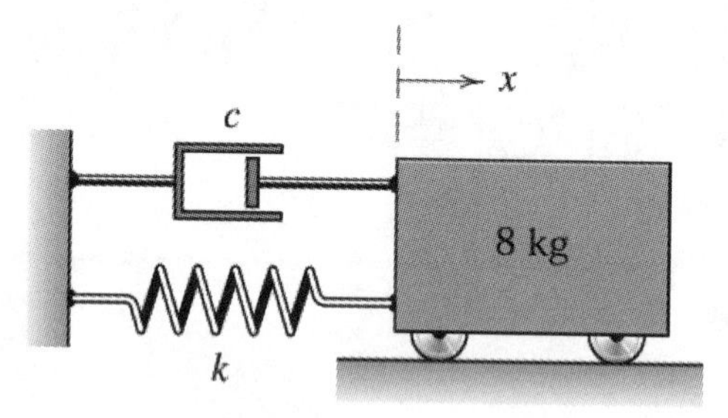

Problem Formulation

As in the sample problem, the natural circular frequency is $\omega_n = \sqrt{k/m} = \sqrt{32/8} = 2$ rad/sec. Now we find the damping ratio for each case (from $\zeta = c/2m\omega_n$) with the result ζ = 0.25, 1, and 1.75 for c = 8, 32, and 56 N•s/m respectively.

(a) ζ = 0.25. Since $\zeta < 1$, the system is underdamped. The damped natural frequency is $\omega_d = \omega_n\sqrt{1-\zeta^2} = 1.937$ rad/sec. The displacement and velocity are

$$x = Ce^{-\zeta\omega_n t}\sin(\omega_d t + \psi) = Ce^{-t/2}\sin(1.937t + \psi)$$

$$\dot{x} = -0.5Ce^{-t/2}\sin(1.937t + \psi) + 1.937Ce^{-t/2}\cos(1.937t + \psi)$$

From the initial conditions $x_0 = 0.2$ and $\dot{x}_0 = 0$ we find C = 0.207 m and ψ= 1.318 rad.

(b) ζ = 1. For ζ = 1, the system is critically damped. The displacement and velocity are

$$x = (A_1 + A_2 t)e^{-\omega_n t} = (A_1 + A_2 t)e^{-2t}$$

$$\dot{x} = A_2 e^{-2t} - 2(A_1 + A_2 t)e^{-2t}$$

From the initial conditions $x_0 = 0.2$ and $\dot{x}_0 = 0$ we find A_1 = 0.2 m and A_2 = 0.4 m/s.

(c) ζ = 1.75. Since $\zeta > 1$, the system is overdamped. The displacement and velocity are

$$x = B_1 e^{\left(-\zeta + \sqrt{\zeta^2 - 1}\right)\omega_n t} + B_2 e^{\left(-\zeta - \sqrt{\zeta^2 - 1}\right)\omega_n t} = B_1 e^{-0.628t} + B_2 e^{-6.372t}$$

$$\dot{x} = -0.628B_1e^{-0.628t} - 6.372B_2e^{-6.372t}$$

From the initial conditions $x_0 = 0.2$ and $\dot{x}_0 = 0$ we find B_1 = 0.222 m and B_2 = –0.0219 m/s.

MATLAB Script

```
%%%%%%%%%%%%%%% Script %%%%%%%%%%%%%%%%%%%%%%
t=0:0.01:5;
C = 0.207; psi = 1.318;
xa = C*sin(1.937*t+psi).*exp(-t/2);
A1 = 0.2; A2 = 0.4;
xb = (A1+A2*t).*exp(-2*t);
B1 = 0.222; B2 = -0.0219;
xc = B1*exp(-0.628*t)+B2*exp(-6.372*t);
x0 = 0.*t;
plot(t,xa,t,xb,t,xc,t,x0)
xlabel('time (sec)')
ylabel('x (m)')
%%%%%%%%%%%% end of script %%%%%%%%%%%%%%%%%%
```

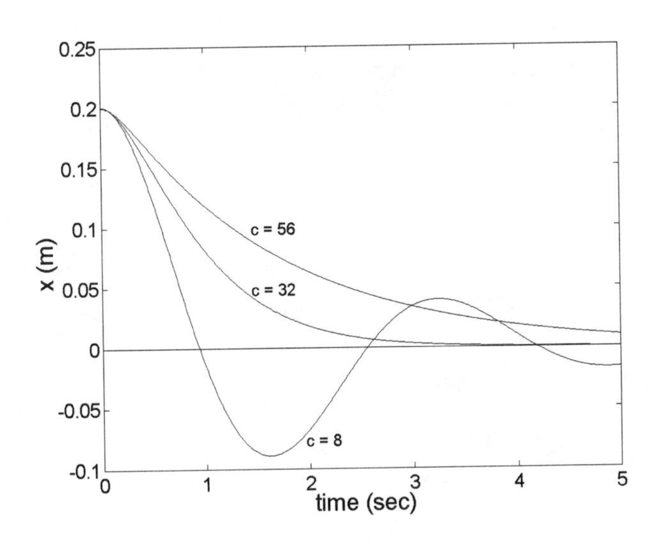

8.2 Sample Problem 8/6 (Forced Vibration of Particles)

The 100-lb piston is supported by a spring of modulus k = 200 lb/in. A dashpot of damping coefficient c = 85 lb-sec/ft acts in parallel with the spring. A fluctuating pressure p = 0.625 sin(30t) (psi) acts on the piston, whose top surface area is 80 in^2. Plot the response of the system for initial conditions x_0 = 0.05 ft and $\dot{x}_0$ = 5, 0, and –5 ft/sec.

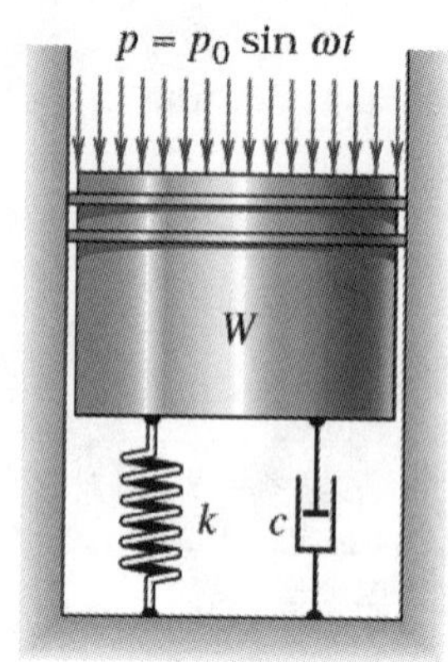

Problem Formulation

The particular (steady state) solution was found in the sample problem in your text,

$$x_p = X \sin(\omega t - \phi)$$

where X = 0.01938 m, ϕ = 1.724 rad and ω = 30 rad/sec. Also from the sample problem, $\omega_n = \sqrt{k/m} = 27.8$ rad/sec and $\zeta = c/2m\omega_n = 0.492$.

The complete solution is found by adding the complementary (transient) and particular solutions. Since the system is underdamped ($\zeta < 1$), the complementary solution is,

$$x_c = Ce^{-\zeta\omega_n t} \sin(\omega_d t + \psi)$$

where $\omega_d = \omega_n\sqrt{1-\zeta^2} = 24.2$ rad/sec. The displacement of the system is

$$x = x_c + x_p = Ce^{-\zeta\omega_n t} \sin(\omega_d t + \psi) + X \sin(\omega t - \phi)$$

$$x = Ce^{-13.68t} \sin(24.2t + \psi) + 0.01938 \sin(30t - 1.724)$$

The velocity is found by differentiating x,

$$\dot{x} = C\omega_d e^{-\zeta\omega_n t} \cos(\omega_d t + \psi) - C\zeta\omega_n e^{-\zeta\omega_n t} \sin(\omega_d t + \psi) + X\omega \cos(\omega t - \phi)$$

$$\dot{x} = Ce^{-13.68t}(24.2\cos(24.2t + \psi) - 13.68\sin(24.2t + \psi)) + 0.5814\cos(30t - 1.724)$$

The constants C and ψ are found from the initial conditions,

$$x_0 = 0.05 = C\sin\psi + X\sin\phi = C\sin\psi - 0.0192$$

$$\dot{x}_0 = C\omega_d\cos\psi - C\zeta\omega_n\sin\psi + X\omega\cos\phi = 24.2C\cos\psi - 13.7C\sin\psi - 0.0887$$

The first equation can be solved for $C = 0.0692/\sin\psi$. Substitution into the second equation gives ψ.

$$\psi = \tan^{-1}\left(\frac{1.675}{\dot{x}_0 + 1.035}\right) \qquad C = \frac{0.0692}{\sin\psi}$$

This yields the following values for C and ψ.

$x_0 = 0.05$ ft, $\dot{x}_0 = 5$ ft/s, $C = 0.259$ ft, $\psi = 0.271$ rad
$x_0 = 0.05$ ft, $\dot{x}_0 = 0$ ft/s, $C = 0.081$ ft, $\psi = 1.017$ rad
$x_0 = 0.05$ ft, $\dot{x}_0 = -5$ ft/s, $C = -0.178$ ft, $\psi = -0.400$ rad

MATLAB Script

```
%%%%%%%%%%%%%%%% Script %%%%%%%%%%%%%%%%%%%%%
% This script plots x versus t for the three
% sets of initial conditions
x = inline('C*sin(24.2*t+psi)*exp(-13.68*t)+.01938*sin(30*t-1.724)')
x = vectorize(x);
t = 0:0.001:0.5;
plot(t,x(.259,.271,t),t,x(.081,1.017,t),t,x(-.178,-.4,t))
xlabel('t (sec)')
ylabel('x (ft)')
%%%%%%%%%%%%% end of script %%%%%%%%%%%%%%%%%%
```

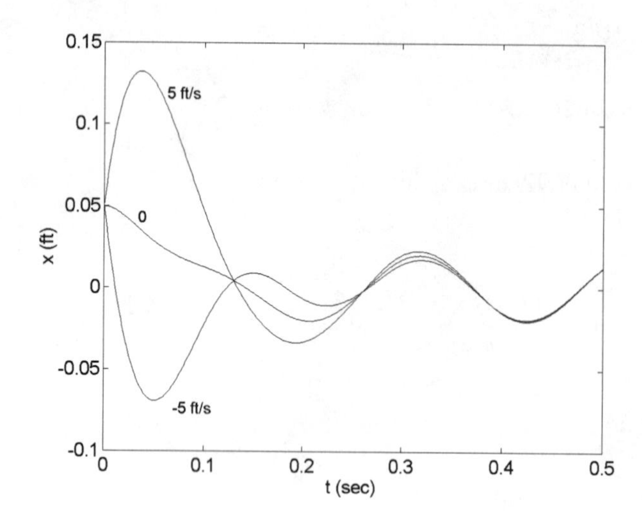

Notice how quickly the three cases converge to the steady state solution.

8.3 Problem 8/137 (Forced Vibration of Particles)

The 10-kg oscillator contains an unbalanced motor whose speed N in revolutions per minute can be varied. The oscillator is restrained in its horizontal motion by a spring of stiffness k = 1080 N/m and by a viscous damper whose piston is resisted by a force of 30 N when moving at a speed of 0.5 m/s. Determine the viscous damping factor ζ and plot the magnification factor M for motor speeds from zero to 300 revolutions per minute. Determine the maximum value of M and the corresponding motor speed.

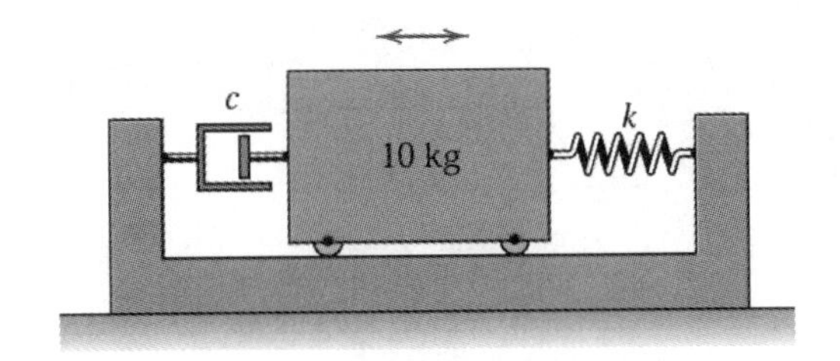

Problem Formulation

Start with the equation for the magnification factor (equation 8/23 in your text)

$$M = \frac{1}{\left\{\left[1-(\omega/\omega_n)^2\right]^2 + \left[2\zeta\omega/\omega_n\right]^2\right\}^{1/2}}$$

The force on a viscous damper is equal to the damping coefficient c times the velocity. Thus, c = 30/0.5 = 60 N-sec/m. We now obtain M in terms of the motor speed N by making the following substitutions,

$$\omega_n = \sqrt{k/m} = \sqrt{1080/10} = 10.39 \text{ rad/s}$$

$$\omega = \frac{2\pi}{60} N$$

$$\zeta = \frac{c}{2m\omega_n} = \frac{60}{2(10)(10.39)} = 0.289$$

The motor speed N where the maximum M occurs is determined by solving the equation dM/dN = 0 for N. The value of N thus determined is substituted back into the expression for M to determine the maximum magnification factor. The result is, M_{max} = 1.809 at N = 90.6 rev/min. Note that this N is somewhat less than the natural frequency 10.39 rad/sec (99.2 rev/min).

MATLAB Worksheet and script

```
%%%%%%%%%%%%%%%% Script %%%%%%%%%%%%%%%%
% This script plots the magnification factor
% M versus N
omega_n = sqrt(1080/10);
c = 30/0.5;
zeta = c/2/10/omega_n;
N = 0:0.1:300;
omega = 2*pi*N/60;
M = 1./sqrt((1-(omega/omega_n).^2).^2+(2*zeta*(omega/omega_n)).^2);
plot(N,M)
xlabel('N (rev/min)')
title('Magnification Factor')
%%%%%%%%%%% end of script %%%%%%%%%%%%%%%%
```

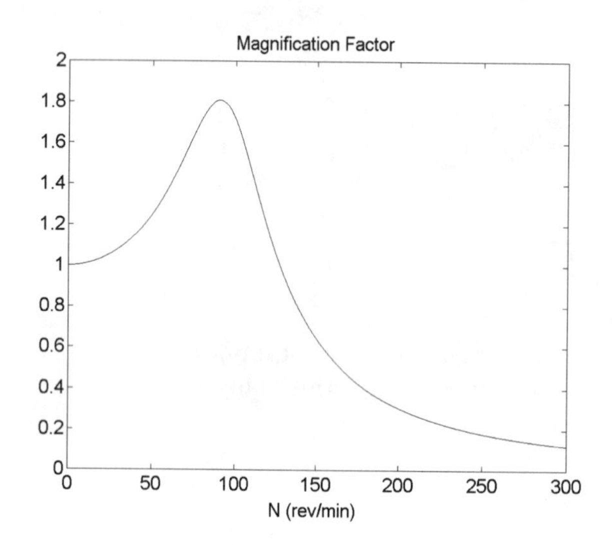

Finding M_{max}

```
EDU» syms N
EDU» omega_n = sqrt(1080/10);
EDU» c = 30/0.5;
EDU» zeta = c/2/10/omega_n;
EDU» omega = 2*pi*N/60;
EDU» M = 1/sqrt((1-(omega/omega_n)^2)^2+(2*zeta*(omega/omega_n))^2)
M =
540/(291600*(1-1/97200*pi^2*N^2)^2+pi^2*N^2)^(1/2)

EDU» dM = diff(M,N)
```

```
dM =
-270/(291600*(1-1/97200*pi^2*N^2)^2+pi^2*N^2)^(3/2)*(-12*(1-
1/97200*pi^2*N^2)*pi^2*N+2*pi^2*N)

EDU» solve(dM,N)
ans =
[            0]
[  90/pi*10^(1/2)]
[ -90/pi*10^(1/2)]

EDU» eval(ans)
ans =
   0
   90.5926
  -90.5926

EDU» subs(M,N,90.5926)
ans  = 1.8091
```

8.4 Problem 8/91 (Rigid Bodies)

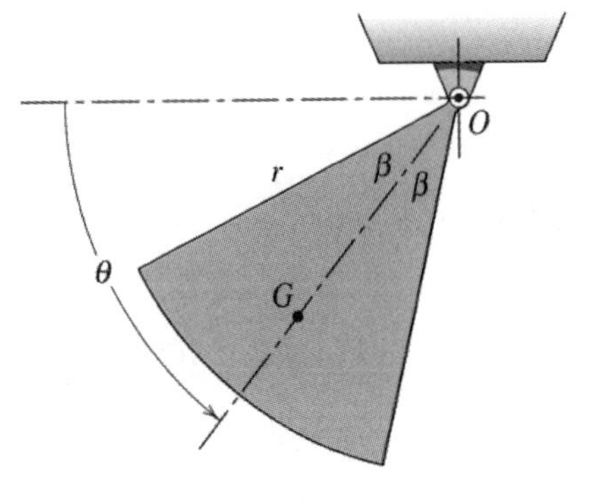

The circular sector of mass m is cut from steel plate of uniform thickness and mounted in a bearing at center O so that it can swing freely in the vertical plane. Determine the natural circular frequency of the sector for small oscillations about the vertical axis ($\theta = \pi/2$). Plot ω_n versus β for β between 0 and 90° assuming that r = 1 m. Specialize your result for ω_n by assuming that β is a small angle and determine the per-cent error E introduced by this assumption. Plot E versus β and find the angle β where the error is 5 %.

Problem Formulation

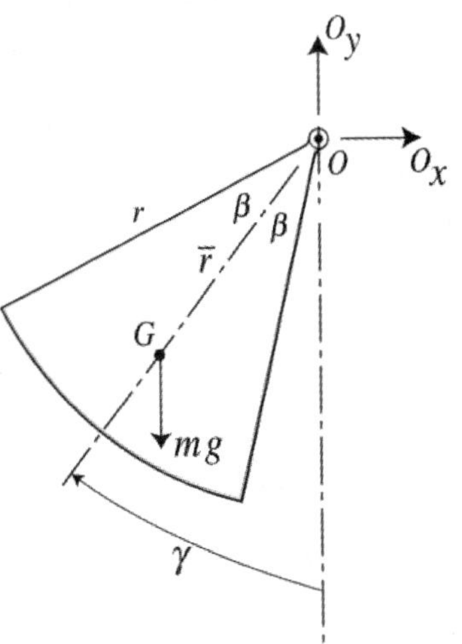

Since the oscillations are about the vertical axis it is much easier to solve this problem in terms of an angle measured from the vertical axis (the angle γ in the free-body diagram shown to the right). It is very important to pay attention to sign conventions. While the angle θ is positive counter-clockwise, γ will be positive clockwise.

$$[\Sigma M_O = I_O \alpha] \qquad - mg\bar{r}\sin\gamma = I_O \ddot{\gamma}$$

From Table D/3 in your text,

$$\bar{r} = \frac{2}{3}\frac{r\sin\beta}{\beta} \qquad I_O = \frac{1}{2}mr^2$$

Substitution of these results and assuming small oscillations ($\sin\gamma \cong \gamma$) yields,

$$\frac{1}{2}mr^2\ddot{\gamma} + \frac{2}{3}\frac{mgr\sin\beta}{\beta}\gamma = 0$$

$$\omega_n = \sqrt{\frac{2mgr\sin\beta/3\beta}{mr^2/2}} = 2\sqrt{\frac{g\sin\beta}{3r\beta}}$$

For small angles β ($\sin\beta \cong \beta$), this result simplifies to

$$\omega_{ns} = 2\sqrt{\frac{g}{3r}}$$

The per-cent error can now be determined as,

$$E = \frac{\omega_{ns} - \omega_n}{\omega_n} \times 100 = 100\left(\sqrt{\frac{\beta}{\sin\beta}} - 1\right)$$

The angle β for which the error is 5 % is determined below to be about 43°. This will undoubtedly be a surprising result for many students. Quite often we say a quantity is small when it can be neglected in comparison with unity. This is not what is generally meant by a small angle. In the present case we mean by small angle that the sine of the angle is roughly equal to the angle (in radians). We will quickly take a look at the errors introduced in the small angle approximations for cosine and tangent in the “for further study” section at the end of this problem.

MATLAB Worksheet and Scripts

```
%%%%%%%%%%%%%% Script #1 %%%%%%%%%%%%%%
% This script plots the natural frequency
% versus beta
r = 1; g = 9.81;
beta = 0:0.01:pi/2;
omega_n = 2*sqrt(g*sin(beta)/3/r./beta);
```

```
plot(beta*180/pi, omega_n)
xlabel('beta (degrees)')
ylabel('natural frequency (rad/s)')
%%%%%%%%%%%%% end of script ************
```

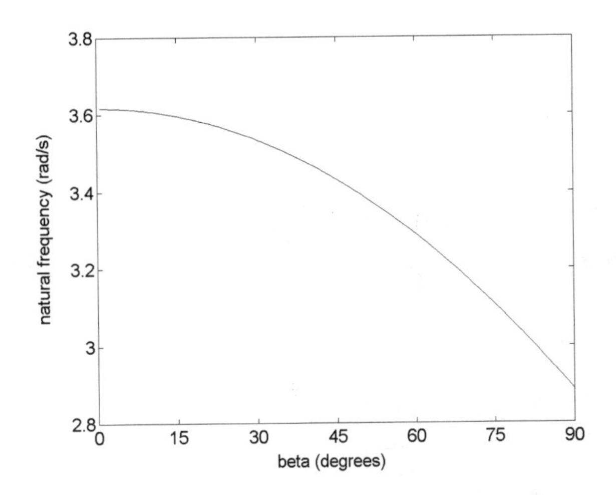


```
%%%%%%%%%%%%%% Script #2 %%%%%%%%%%%%%%
% This script plots the per-cent error
% versus beta
beta = 0:0.01:pi/2;
E = 100*(sqrt(beta./sin(beta))-1);
plot(beta*180/pi, E)
xlabel('beta (degrees)')
ylabel('per-cent error')
%%%%%%%%%%%%% end of script %%%%%%%%%%%
```

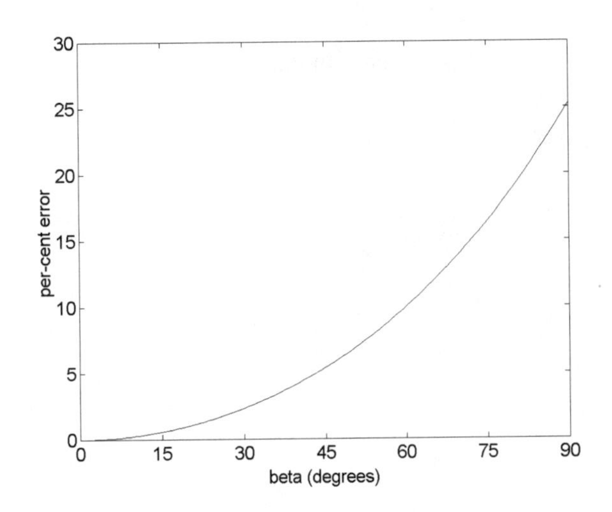

Now we want to find the angle β for which $E = 5$ %. Our method of choice here would be to use the symbolic *solve* command to solve the equation $E - 5 = 0$. Occasionally this approach fails to find the solution. Unfortunately, the present problem is one of those cases. So, we set up a function m-file and use *fzero*.

```
%%%%%% Function m-file %%%%%%
function y = E(x)
y = 100*(sqrt(x./sin(x))-1)-5;
```

EDU» fzero('E',0.1)
Zero found in the interval: [-0.62408, 0.82408].

ans = 0.7577

EDU» ans*180/pi
ans = 43.4118

For Further Study

Investigate and compare the per-cent error introduced by each of the following small angle approximations.

$$\sin\theta \cong \theta$$

$$\cos\theta \cong 1$$

$$\tan\theta = \sin\theta / \cos\theta \cong \theta$$

A per-cent error can be defined as above for each case and then plotted versus θ.

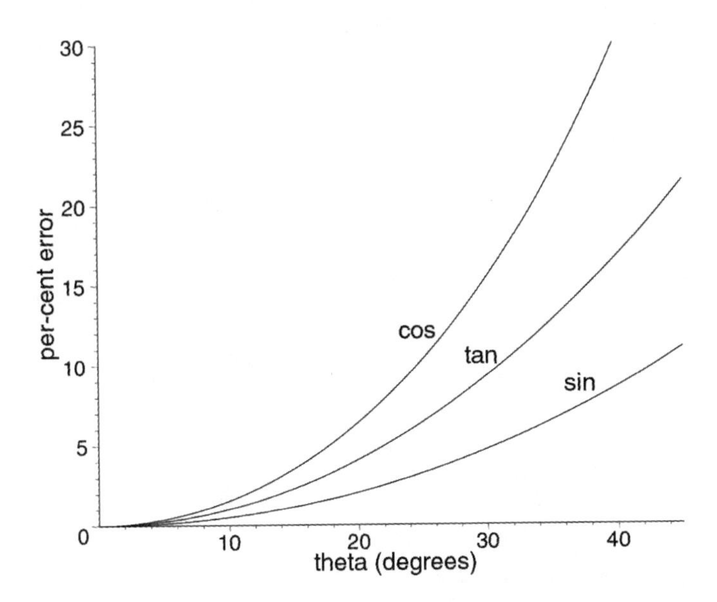

Clearly cosine is the worst case. There is a 5 % error in the approximation for cosine when $\theta = 17.8°$.